NURSING MANAGERS POWERED INCREDIBLY CLOSE OVERVIEW

Nursing Managers Powered Incredibly Close Overview

Shreen Gaber
CCLN, PhD
Nursing Administration Department
Faculty of Nursing
Cairo University

Yale | University Press

2018

First Printing: 2018

ISBN: 978-0-359-31605-2

Yale University Press

Yale University Press
New Haven,
CT 06520-9040
P.O. Box 209040
Telephone: 1-800-405-1619
Connecticut
USA
www.yalebooks.yale.edu

U.S. trade bookstores and wholesalers: Please contact;
Shreen Gaber Tel: (+20)100-8144971
Cairo University
Fax: (+20) 23657190 or email: Sameh17@cu.edu.eg

Shreen Gaber

To my lovely huspend and children

Thank you. Without your support and persistence, I would have

never accomplished this work.

 # Content

 # Acknowledgements

I would like to express my great thanks and appreciation to my family, parents, colleagues, teachers at Faculty of Nursing-Cairo University who are always willing to provide their support and guidance. I also appreciate the efforts of (Yale University Press) especially Dr. Ann Kurth for guidance and editing acquisition.

Shreen Gaber

 **Preface**

Empowerment of nurse leaders is vital in enabling nursing teams to deliver high-quality care. Administrative support is needed to counterpoise the disempowering impact of financial, resource pressures and sustain practice engagement. This book allurements and attention to the different type, sources, barrier of leaders' power and how to enhance it. Implications for Nursing Management: Empowered nurse managers at all levels who feel reinforced by their institutions are more likely to stay in their roles, remain committed to achieving quality patient care and act as influential persuasive role models for prospective future leaders. Whether you are a novice or qualified health care provider, I faith that you find this book enjoyable and clinically crucial.

 **Nursing Managers Powered Incredibly Close Overview**

The Concept of Empowerment

Empowerment is not a new word in the English language. According to the Oxford English Dictionary (Simpson & Bradley 1989) the words empowering and empowering were used as early as the 17th century in administrative letters. The Dictionary's definition of the verb empower is `to invest legally or formally with power or authority; authorize, license'. Empowerment is defined as `the action of empowering, the state of being empowered'. The Etymological Dictionary (Partridge 1966) leads us back to the root of power. The Latin word potter means to be able and to have the ability to choose. Synonyms for empowerment listed in The Original Roget's Thesaurus (Kirkpatrick 1992) include `make possible, commission, permit, invest with power, authorize, allow, facilitate', and for empowered `powerful, authoritative'. According to these definitions, the concept may be understood both in terms of qualities and as a process associated with the individual and the environment (Simpson & Bradley 1989).

Empowerment is an abstract concept that is fundamentally positive, referring to solutions rather than to problems. It is also a dynamic concept: power is taken over and given away, power is shared. In the empowering process individuals, organizations and communities pursue maximal impact on their own life and eventual choices (Kieffer 1984, Gibson 2017). Empowerment is associated with growth and development. The process of individual empowerment requires critical introspection and changing patterns of activity accordingly. At the community level empowerment is understood in terms of people uniting to achieve common goals. Empowerment has also been described in terms of a negation, i.e. through its absence. Since human decisions differ from person to person and are individual within organizations, it is difficult to give a brief and unambiguous definition of the concept. (Rappaport 1984, Rodwell 1996).

The Crux of Leader Power

In order to understand properly the meaning of empowerment we also need to look at the concept of power. In the nursing context, the word power usually has negative connotations: it is associated with hierarchical organization and authoritative leadership, with one person restricting another's freedom of action. In critical social theory in particular, power is interpreted in terms of coercion and domination. Power is extra personal, which means that an increase in power has to be compensated by someone else surrendering part of their power. This process of relinquishing power has also been called legitimation. According to this view legitimate power includes power associated with standing, action and expertise. Power may also be connected with knowledge, coercion, or it may be conditioned (Sheldon & Parker 1997)

Kanter (1979) states that, instead of coercion and domination, power could mean efficacy and goal-orientation. Organizations do not in themselves prohibit or provide power, but it is generated by each individual through their own personal actions. Neither does power necessarily form part of a hierarchical system so that those at a higher level dispose of more power than those at lower levels. Even leaders may lack power. The key tools which help to generate power are the creation of opportunities, effective information and support at each level of the organization. Robbins (1986) describes successful leadership with similar qualities: strong confidence, unambiguous strategies and values, vigor, communicative capacities and a uniting, committing power. Robbins' concepts give the impression of almost ideal leadership and collegiality.

The empowerment of staff is hallmark of transformational leadership. Empowerment means to enable, develop, or allow. Hawks (1992) defines empowerment as an interactive process that develops, builds, and increases power cooperation, sharing, and working together. Strader and Decker (1995) define empowerment as the process by which a manager or leader shares power with other.

Empowerment plants seeds of leadership, colleagueship, self-respect, and professionalism. In addition, it frees staff from mechanistic when thinking and encourages critical thinking, problem solving, and the application of knowledge to practice. Nurses are empowered when administrators and managers share authority with them. Nurses seek community with other nurses as a form of empowerment. Their power is extended by new technologies and the ability to use them. They are empowered at work by computers with modems, cellular phones, fax machines, and access to e-mail systems. Nurses are empowered when society rewards their initiative as individuals. Empowerment is a management approach designed to give frontline employees the authority they need to do what needs to be done without having to check with management. In spite of the entire favorable buzz, there is little hard evidence that it has really made much difference in routine organizational life. Some empowerment does exist and, when accompanied by accountability and appropriate guidance, it can lead to increased employee and customer satisfaction.Significant employee empowerment is rare, and it is not easy to initiate or maintain.

Concept of Power

Ability to get things done, to mobilize resources (Kanter, 1993). Influencing decisions, controlling resources, and affecting behavior. Nurses need power to be able to influence patients, physicians, and other care professionals, as well as each other. that force that enables persons or groups to realize their will even against opposition." Nurses become empowered through education, leadership, and collective action. Power in nursing is based on four facets: expertise and reputation, position or profession, personality, connections to influential people, such as major corporations, organizations, and politicians, so empowered nurses are: Highly motivated (lashinger & havens, 1996). Able to motivate and empower others by sharing the Sources of power (lashinger & havens, 1996, Less burnout (lashinger, finagan, 2003(, less job strain (lashinger, shamian, 2001)

4

Lashinger determined the Managers Powerlessness occurs when nurses:

- Are not recognized.
- Are not appreciated.
- Are not paid attention.
- …what happens⁇⁇
- They don't feel they can make an impact.
- They will not be creative.
- Teams feel they can't make change happen.
- Negative organizational climate.
- Frustration

Types of Power

Formal power which results from jobs that afford flexibility, visibility and are relevant to key organizational process Informal power evolves from an individual's network of alliances with sponsors, peers and subordinates both within and outside of the organization"...

Power can be classified into:

A- Personal power: (expert, referent and connection, persuasive power.(

B- Positional power: (reward, punishment and legitimate power.(

Sources of power

There are several sources for exercising power

(1)   Reward power:is power obtained by the ability to grant favors & reward others with whatever they value . The uses of rewards foster a great deal of loyalty of the subordinates.

(2)   Punishment or coercive power:is the opposite of reward power , it is based on fear of punishment if the expectations of manager are not met . The manager can obtain compliance threats of transfer, dismissal or ignores the employees.

(3)   Legitimate power:is a formal position power , also authority is a legitimate power . It stems from the right of the manager to make a request because of the authority associated with job or rank in an organizational hierarchy.

(4)   Expert power:is power gained through knowledge , expertise or experience . Possessing required knowledge, skills and competence allows a manager to gain power over others who need that knowledge, skills and competence.

(5)   information power:Arises from the ability to access and share information

(6)   Referent power: Informal power comes from others recognizing that individual has special qualities and is admired .An individual may develop referent power because others perceive him as powerful. This perception based on personal charisma (the way the individuals speaks or acts). Referent power is used by those who are accepted by others as role model or ideals.

(7)   Connection power :is based on an individual's formal and informal links to influence persons within and outside an area or organization . It relates to the status and visibility of the individual as well as the position .

(8)   Persuasive power: influencing others by providing an effective point of view or argument (kinner, 2010). A postmodern view on the concept of power is offered by the French philosopher

Michel Foucault. Foucault suggests that the world is in itself complex, divergent and polymorphic. Power is omnipresent because it originates from everywhere. It is difficult to personify power because it is involved in all human interaction. Power and knowledge are closely interwoven: where there is power there is also knowledge, and power begets knowledge. Similarly, power increases through knowledge. Man is at once both the object and the subject in relation to power. In Foucault's view power is not exclusively legalized power, decision-making, or technique.

The essence of power cannot be grasped simply by asking who has access to power or what the implications of power are. The exposure of power must be prospected at the individual level. People are the means and the generators of the power exercised. It follows that a study of the essence of power must be based on a bottom-up rather than top-down analysis. It is often presumed that power is a logical, causal relationship; we keep talking about taking over and retaining power. The real tenant of power has taken up a position at the right junction of human relations. The exercise of power is not so much action, domination or control but the real exercise of power turns out to consist of the manipulation of thoughts, attitudes and social relationships. The disposal of power also requires assuming some responsibility. Herein lies the reciprocal essence of power relationships because the object also retains power towards its exerciser. It is difficult to describe power and its exercise simply in terms of a causal relation, ordinance or delegation. Rather, it consists of a multilayered network of social relations requiring consciousness of basic human desires and

Importance of manager's empowerment

The global importance of nurse empowerment has become apparent especially the in studies which have results that confirmed a positive influence on organizational outcomes, notably patient safety only and reduction of falls. Influences on workforce outcomes have also demonstrated an impact on stress, burnout, civility, trust, organizational commitment, job satisfaction and intention to leave (Cicolini et al. 2014, Wang et al. 2014). In a survey conducted in 12 countries in Europe and the USA, concluded that all hospitals sampled faced problems of quality, safety, and nurse burnout and job dissatisfaction. Implementation of measures to contain financial costs and improve quality in the work environment have impacted variably on nurse staffing, workloads and skill-mix (Aitken et al. 2012)

Empowerment is giving or delegation of power and authority and it construct used to maximize organizational efficacy. More, organizational nurse managers empowerment has been defined as having access to information, support, resources and opportunity, which enables nurse leaders to be most effective and positive about their work and when accessed are sources of power. In difference, Oshry 2016) to achieve mutual enhancement he views systems power in organizations as focused on enabling survivability and development, and the potential power of groups and individuals in the middle condition is to integrate systems components. Psychological empowerment is defined as motivational, containing elements of meaning, competence, self-determination, impact, and self-efficacy and is also characterized by a sense of perceived control and achieving goals.

The majority of a systematic review has confirmed that manufacture of an environment which contributes to structural empowerment of nurses also contributes to psychological empowerment, leading to positive work behaviors and attitudes (Wagner et al. 2012). For nurses especially in the middle management positions, leaders' empowerment is of vital importance, given their influence on clinical practice, exerted through leadership roles enabling nursing teams to deliver high-quality care. More widely, transformational leadership behaviors of all leaders, chief executives, middle and frontline nurse managers can also impact on the empowerment of subordinate staff, enhancing their job satisfaction through the most vital factor which is enhancement of trust between them (Purdy et al. 2010).

Perceptible benefits such as :

1) It is much easier to find the best solution to a business problem،

2) The diverse ideas are shared and implemented ،

3) The decisions are made at lowest level of the organization ،

4) The workgroup is recognized for its efforts and performance،

5) The individual have the opportunity to influence the goals of the workgroup،

6) The acquisition of new skills and knowledge are encouraged and facilitated ،

7) The organization structure becomes flattened and less hierarchical‹

8) The managers have more time to lead and not to micromanage employees' activities‹

9) This provides meaningful, measurable positive business impact.

The incorporeal benefits of empowerment such as :

1) It allows leveraging the collective strengths of all group members ‹

2) The group takes the ownership of the statement of work and results‹

3) It helps to create a culture of trust and collaboration ‹

4) It enhances the individual self-esteem ‹

5) It improves the communication among the mangers and the employees‹

6) It provides a more enjoyable working environment.

Empowerment theoretical framework

Kanter's theoretical framework (1993) developed in a context identified structure of power, opportunity and proportional distribution as determinants of organizational structural empowerment. Each has related elements which are either empowering or disempowering depending on enactment, notably role models, hierarchies, resources, commitment, quality, relationships and team work. Individuals obtain formal power as part of their roles, in which elements of power are encapsulated in the job description, title and closely linked to the organization's culture, vision and freedom to make decisions. Informal power is obtained through net-works and staff alliances that allow influence to be exerted and desired outcomes achieved. Access to opportunity empowers by enhancing the development of knowledge and skills through personal and professional development, enabling promotion. Proportional distribution, refers to the number of people with similar roles and attributes; those who are represented in large proportions are seen as members of a group, finding it easier to gain credibility and access networks. Subsequent investigations over the last two decades, predominately in Canada and the USA, have developed, expanded and applied Kanter's theoretical framework to nursing, identifying access to resources, opportunity, information, and support, formal and informal power as key elements of structural empowerment (Laschinger et al. 2004).

In distinction, Oshry (2007) developed a total systems approach to power in organizations based on empirical studies, which is of particular relevance to empowerment and disempowerment experienced by middle managers. This approach views organizations as social systems and subsystems existing in environments posing threats and opportunities. System power is defined as the power each part of the system has to enhance for whole system survivability and development and group empowerment is conceptualized as supporting individual empowerment. Furthermore, he postulates that whatever an individual's position in the organizational hierarchy, movement between three core conditions is experienced to a variable degree, top (overall responsibility for system, initiatives, resource control), middle (overall responsibility for team management, mediation of initiatives, no direct resource control) and front lines (on receiving end of initiatives, no resource control.

Kanter's (1993) theory of structural empowerment consists of four components: access to opportunity, learning and career advancement; access to information, necessary knowledge for the work; access to sup-port from managers and subordinates; and access to resources, such as time, money and supplies. Formal and informal powers are components thought to influence an individual's access to structural empowerment. Having access to all these components is the most important factor influencing an individual's working life (Kanter 1993). In research (Laschinger et al. 2010) on nursing leaders, senior nurse leaders perceived their working life to be better than that of middle managers, who in turn perceived their working life to be better than that of FLMs. Moreover, senior nurse leaders perceived higher structural empower-ment and higher job satisfaction than did middle managers and FLMs (Laschinger et al. 2010). Further-more, in studies of nurse managers (Laschinger et al. 2007) and FLMs (Abdelrazek et al. 2010), positive relationships have been revealed between structural empowerment and psychological empowerment. Spreitzer (1995) defines psychological empowerment as an individual's reflections on their work role and as consisting of four cognitive dimensions: meaning, the value of the workplace's goals in relation to the individual's ideals; competence, the individual's belief in his/her ability to perform the work with skill and mastery;

12

self-determination, the individual's sense of having autonomy; and control over the work process and impact, the individual's degree of influence over administrative, strategic and operating outcomes at work.

Critical Social Theory and Empowerment

Critical social theory is often associated with improving the living conditions of the underprivileged (Ward & Mullender 1991). Where the individual nurse's professional growth and development is concerned, the perspective must be broadened to encompass the development of the whole profession and its relative position. For instance, an emancipatory starting-point has been suggested for the education of nurses (Harden 1996). Some writers have drawn attention to a transfer effect where an empowered nurse will also inspire increased empowerment among patients (Chandler 1992). Most articles deal with critical social theory and empowerment from a theoretical vantage-point (Parker & McFarlane 1991).

Barriers of leaders' empowerment

- Organization beliefs about authority and status. Empowerment is blocked if authority and power are viewed as the key motivational forces for achieving the organization's mission and strategic planning.
- Control perception, needs, and attitudes. If managers emphasize rules, regulations, mandated policies, and procedures, little room is left for employee participation and empowerment.
- Organizational inertia. Empowerment does not happen "naturally". It occurs only as a result of an organizational commitment of time, energy, and resources .

- Personal and interdepartmental barriers. Interdepartmental rivalries result in internal competition for resources. The more time managers must spend "defending their turf," the less organizational emphasis will be given to the empowerment process .

- Employee number, mix, and skill. Larger organizations with greater staff diversity face a greater challenge in developing focused, yet flexible, strategies to empower their work force .

- A lack of ability and unwillingness of staff to assume responsibility and accountability for their attitudes and behaviors. Clarity of job roles or job expectations encourages empowerment as staff understands what expected of them and can identify areas for improvement .

- Managerial incompetence. The management skills required to empower staff are planning and goal setting, identifying and addressing problems, making decisions, defining priorities, implementing and managing change, forming interactive and self-directed teams, communicating, resolving conflict, fostering motivation, and building consensus.

The feminist movement has tended to associate empowerment primarily with the exercise of power. The development of nursing care and nurses' professional competencies are seen as an exercise in promoting the position of women. This theory suggests that early head nurses were feminists. Some writers point out that efforts to strengthen the professional status of nurses may have adverse effects on the patient's position. This view refers to the traditional relationship between the dominant expert and passive patient (Roberts 1983, Parker & McFarlane 1991, Porter 1994, Huntington et al. 1996, Cheek & Porter 1997). However, critical theory insists that patients also need to be empowered. A therapeutic nurse± patient relationship is implied, based upon mutual respect, trust and equality of worth. Patients must be active and equal participants in their own empowerment. (McNay 2017).

In other words, rather than empower patients, nurses facilitate the empowerment of patients (Rafael 1996, Anderson 1996. (Roberts, et.al.1983) describes nurses as an oppressed group because they have assimilated their values from nursing and because they have adapted their behavior accordingly. The views of the nursing profession as an oppressed group are largely shaped and influenced by myths and beliefs (Harden 1996, Clifford 1992). Most typically, the main obstacles to nurses' independence in hospitals are represented by head nurses and doctors as well as their patriarchal and authoritarian leadership style (Clifford 1992).

Principles of Empowerment

Implementing principles of empowerment can be challenging because it involves a radical shift from our traditional way of operating. The following principles include the most important elements for creating an empowered organization (Laschinger, 2006)

1) People are an organization's most valuable resource.

The founding principle of empowerment is that people are more important than management systems. The essence of this principle is that the manner in which a management system operates is determined by the people who comprise the organization. It assumes that people are not expendable, simply because they bring differences which may force the system of operation to change. Although projects may come and go, the most vital recyclable resource, which is utilized over and over again, is people. For this reason, it is necessary to preserve the mental, physical, emotional, and even the spiritual well-being of employees. In the present progression from the information era to the knowledge-based era to the era of spirituality, the development, utilization, and retention of creative and innovative employees will determine the survival of an organization. (Laschinger, 2006).

2) High-involvement is maximized.

High-involvement is based upon the assumption that the more employees are involved in designing and controlling their work functions, the more productively and efficiently the organization will operate. The basis of this assumption is that structured management systems severely limit the performance capacities of employees. For high-involvement to work, employees must assume responsibility and accountability for understanding and ensuring the successful production of a whole aspect of work. Individually and collectively, employees must have a high degree of self-discipline and self-management in order to operate with the least amount of oversight or management. The crucial fact to understand is that in today's hyper-accelerated world, high-involvement is inevitable. (Schlesinger& Oshry 1984)

3) Teamwork is valued and rewarded.

Teamwork is essential for the success of empowerment since so few products and services can be delivered today by the efforts of a single employee. When this principle is applied to an organization, the organization is viewed as a network of interdependent centers of excellence. The commitment to team projects must be balanced with commitment to the overall success of the organization.

4) Personal and professional growth is continuous.

Personal growth and professional development are a way of life in empowered organizations. Since empowerment is a dynamic process rather than a specific goal to be reached, there is the sequential cycle of self-motivated goal setting and achievement, which continually drives the enhanced capability of employees. High-involvement necessitates people-oriented skills, hence the corresponding necessity for continual personal growth. The most severe limitation to performance in high-involvement organizations is employees' reluctance to proactively accept the process of personal growth. This principle also makes a job interesting, fun, and creative because of the necessity for continuous improvement. (Laschinger, 2006)

5) Responsibility and accountability are maximized.

Empowerment is based upon maximizing individual and collective responsibility and accountability. This means a predisposed mindset of total responsibility for projects or tasks that are delegated. Such a mind-set has the potential for not only meeting but also exceeding customer or client expectations. Without a critical fraction of highly responsible and accountable employees, empowerment is not possible. The more difficult of these two requirements is holding self and others accountable. Accountability is probably the limiting factor in determining the extent to which high-involvement is possible. (Schlesinger& Oshry 1984)

6) Self-determination, self-motivation, and self-management are expected.

An inherent assumption of empowerment is that most, if not all, employees have the talent and capability to perform their jobs and responsibilities with the least amount of oversight and management. The major limitation in fully living up to one's potential is a mindset that compromises self-determination in difficult situations. Where the talent or capability is lacking, principle (4) above applies. An additional assumption is that the incentive to meet and possibly exceed job expectations comes from within an individual. Given principles (4) and (5) above and a clear organizational support system, employees are expected to be self-driven in terms of determination, motivation, and management. (Laschinger, 2006)

7) Expanded delegation is a continual process.

It is vitally important to understand that the act of delegation is not empowerment. The procedure for implementing empowerment is delegation of responsibility within clearly defined guidelines. Empowerment ultimately depends most on an individual's ability to perform the expanded responsibility that has been delegated. A requirement of expanded delegation is ensuring that an individual or a team is maximally prepared to accept the expanded responsibility. A central issue to the success of empowerment is giving up control. This act requires trust and the willingness to share information, knowledge, and power. Also, the mentoring and coaching become critical management skills in support of delegation. Teams and learning pairs provide natural opportunities for mentoring and coaching relationships. (Schlesinger& Oshry 1984)

8) Hierarchy is minimized.

A natural consequence of extensive delegation is the systematic reduction of hierarchy. Hierarchical organizational structures discourage empowerment by supporting a line-of-authority system and discouraging cross-functional teams. Cross-functional teams focus on clients or customers, products, goods, or services. This principle, in an indirect way, means that influence and authority are based on demonstrated competence, performance, and an ability to manage oneself.

Challenges of managers' empowerment

More widely, Schlesinger and Schlesinger& Oshry (1984) identified challenges which can arise for all managers through engagement in certain quality of work life activities (e.g. quality control) which can exacerbate existing problems of inadequate recognition, hectic work pace and lack of influence. Empowering nurse manager's entrée to resources specifically for administrative support is also crucial to facilitating their visibility and work with quality issues (Hagerman et al. 2015). A factor which has implications for nurse leaders in management levels relates to concerns that continual restructuring, a focus on economic outputs, varying levels of organizational visibility, inclusion and lack of financial control experienced by senior nurse managers can impact negatively on their empowerment (While Manojlavich, 2005), found that nurse leaders who work closely with qualified nurses may not always have the power to change practice environments due to hierarchical position, other studies have concluded that empowered middle-level nurse managers can act as influential role models, enhancing access to power, opportunities and resources, enabling nurses to expand their scope of practice.

Organizational and Management Theories

Organization and management theories have made more and more use of the concept of empowerment during the past decade. The number of studies concerned with the development of the nursing organization has also increased. Most of the research has been published in North America (Chandler 1991, Laschinger 1996). The results indicate that the factors involved in empowerment show a positive correlation with staff well-being and commitment to their work. In most cases empowerment in the organizational environment is described as a process. The crucial difference in comparison with social and emancipatory theories is that organization theories do not account for oppressed groups. Power is most frequently associated with standing and it is distributed within the organizational hierarchy from the top down. Empowerment in organizations leads to increased productivity and effectiveness. The means applied include the reorganization of work and the development of human resources management (Clifford 1992).

Social Psychological Theories

The social psychological theory of development describes empowerment from the point of view of the individual. The earliest studies from the 1980s focus on the content of the concept (Hess 1984, Rappaport 1984). More recently, Conger & Kanungo (1988) and Thomas & Velthouse (1990) worked on developing the theoretical background of empowerment. Empowerment is seen as a process of personal growth and development in which key factors are the individuals' characteristics such as beliefs, views, values, perceptions, and relationships with the environment. Environmental variables at the individual level include race, sex, leisure interests, roles and standing. People do not usually pay very much attention to, or take Integrative literature reviews and meta-analyses power and empowerment the trouble to gather, information that is not directly relevant to them. Rejection is nevertheless a precondition for increased knowledge and personal development (Bandura 1978).

Kieffer (1984) and Rappaport (1984) were among the first to describe empowerment as a development process. Kieffer identified three dimensions which enhance the empowerment experience: a positive self-identity, extensive apprehension and rejection of one's environment, and the capability of social intercourse. An empowered person does not pretend to have acquired more power but feels empowered. Power is both surrendered and conquered. The process is dynamic and synergetic and contains both positive and negative elements. It may also prove very painful (Hess 1984 pp. 227±237). In Rappaport's (1984) view empowerment is a process, a `mechanism by which people, organizations, and communities gain mastery over their own lives.

21

Conger & Kanungo (1988) developed the concept of empowerment further from the perspective of motivation theory.

Hardiness in different level of management:

The propensity to experience hardiness is linked to structural empowerment. Therefore, employers should endorse changes in the structural conditions of the working environment. Hardiness modifies the influence of structural empowerment on psychological empowerment. Hardiness helps taking better advantage of the conditions of structural empowerment. Therefore, employers should promote training programs on hardiness. Psychological empowerment facilitates the relationship between structural empowerment and hardiness. A moderate or high level of hardiness is needed for structural empowerment to increase psychological empowerment and to lessen also burnout symptoms that may arise. In the case of low hardiness in different levels of managers, organizations must develop well defined hardiness before enhancing structural empowerment (Sabiston & Laschinger 2015).

Laschinger et al. (2003) have recommended that hardiness can act as a moderating variable in the relationship between the SE and PE and help reduce the impact of the nature of the environment on subsequent burnout. Therefore, the principal purpose of this study is to investigate the role that hardiness could play in moderating the relationship between SE and PE to predict burnout. In so doing, we hope to improve the previous literature on burnout and PE on several points. In the first place, investigating upon the moderating role of hardiness in the relation between structural and PE will contribute to diminish the existing gap in the literature regarding the impact which personality variables have on PE. In the second place, by questioning whether the mediating role of PE in the relation between SE and burnout may be conditioned by hardiness, our work contributes to the improvement of the understanding of burnout. Hardiness can positively affect burnout in different level of managements. In our work, this relationship has been tested on a specifically sample of middle managers who do not work in health care-related settings. Specifically, the sample is composed of Spanish middle managers who worked in different industries, mainly in manufacturing ones (Sabiston & Laschinger 2015).

Burnout and managers: changing from structural empowerment to burnout

Burnout has been identified as one of the most vital emerging hazards/ risks in the current job outlook and, consequently, as one of the major encounters for occupational safety and health faced by European organizations. Among the consequences of this syndrome, the following have been specified: moderated to low job performance; desire to leave a job; extra days of absenteeism; psychological disorders (such as anger, depression, anxiety, and apathy). According to the data of European Agency for Safety and Health at Work (2014), the annual cost of burnout is estimated at over 600 billion euros: €272 billion from absenteeism, €242 billion from loss of productivity, €63 billion from health care costs, and €39 billion from social welfare costs in the form of disability benefit payments. To prevent these harmful consequences, and as far as possible avoid them, it is important to know what factors contribute to its occurrence. The work environment is one of the most important antecedents of burnout showed in most studied by different researchers (Van Bogaert, Kowalski, Weeks, Van Heusden, & Clarke, 2013).

Previous research has shown that structural empowerment approves the work efficiency and prevents burnout Kanter's theory argues that access to information, resources, support, and the opportunity to learn and develop are the organizational empowerment structures that enhance employees' power to accomplish work. Currently, several researchers (Spreitzer, 2008) consider structured empowerment as an antecedent to psychological empowerment (PE). This is what defined as an 'intrinsic task motivation reflecting a sense of control in relation to one's work and an active orientation to one's work role'. Studies also showed that SE (time 1) has a direct effect on PE (time 1) and an indirect effect on burnout (time 2) through PE. This mediating role of psychological empowerment in the relation between structured empowerment and strain outcomes has been authenticated by important reviews and meta-analysis (Maynard, Gilson, & Mathieu, 2012).

Managers' role/ duties

The managers' responsibilities are often strategic, long term and focused on maintaining their subordinates' empowerment. Currently, organizations have become flatter, more global, and lighter, which obliges the middle managers to do more with fewer resources while they face uncertainty caused by the rapid and continuous changes occurring in the market (McKinney,

McMahon, & Walsh, 2013). With much pressure, many responsibilities and little economic and staffing resources, middle managers are required to exercise leadership in difficult conditions and it is no phenomenon that many are suffering increasing levels of burnout. The stressful work nature of the middle managers' role specifically could have negative consequences on their physical and psychological health as on the appropriate performance of their work. If a middle manager experiences burnout and feels demoralized, this will result in difficulties encouraging his/her subordinates to use new approaches or supporting them in the development of their personal potential (Judkins, Furlow, & Kendricks, 2007).

Yet, surprisingly, there are very few authors who have studied the impact of structural and psychological empowerment on the burnout of different level of managements. The only work we have knowledge of is by Almost et al. (2004), who based themselves on a sample of nurse managers. Which results showed that when structured empowerment and psychological empowerment are used as independent predictors of burnout, structured empowerment becomes the sole significant predictor. However, an interaction analysis would allow for both structured empowerment to contribute to explaining the burnout in middle managers. It seems thus essential to expand the investigations concerning how structured empowerment influences the explanation of burnout in middle managers. More concretely, it is interesting to expand the empirical research to determine how structured empowerment, psychological empowerment.

In Spain, the works on structural and/or psychological empowerment in manufacturing businesses have mostly centered upon exploring: (1) which are the high-involvement work practices (high-performance work practices or high-commitment work practices) used by companies to develop the concession and involvement of employees and to obtain a broader identification with the goals of the firm, (2) which factors may have an influence on the implementation of high-involvement work practices in firms and (3) how high-involvement work practices influence on quality and time-based performance, on production outcomes (Marin-Garcia & Bonavia, 2015), or on operational performance (cost, quality, flexibility, and delivery) and financial performance (effectiveness) (Vazquez-Bustelo & Avella, 2017).

For our knowledge, there are no previous works having studied, within the very context of the manufacturing industry and the specific samples of middle managers, how structural and/or psychological empowerment, and hardiness contribute to predicting burnout. Therefore, our contribution in this study aims at improving the existing literature on burnout and the SE and PE in middle managers, which has to date hardly no published studies. Added to that, our work aims at amplifying the existing literature on structural and/or psychological empowerment, and burnout within the Spanish manufacturing industry. Therefore, the results of this investigation could be important for (1) keeping the physical and mental health of middle managers, which tends to deteriorate with burnout and (2) retaining middle managers, thereby lowering turnover and absenteeism, which will have a positive impact on the company's bottom line.

The relationship between empowerment and burnout

Burnout 'is a response to chronic emotional and interpersonal stressors on the job' (Maslach, Schaufeli, & Leiter, 2001). Burnout arises as a result of increasing emotional exhaustion, cynicism, and lack of professional efficacy (Salanova, Schaufeli, Llorens, Peiro, & Grau, 2000). Emotional exhaustion refers to the draining of personal resources, together with the feeling that one no longer has anything to offer others psychologically. Cynicism refers, in general terms, to attitudes of indifference regarding the work carried out by the subject. Lastly, lack of professional efficacy is the feeling of a lack of effectiveness in the workplace. Burnout is related to the work and the role that the person performs in their work (Lim, Bogossian, & Ahern, 2010). The researchers seem to agree in pointing to factors related to work as being key in the development of burnout. Burnout is especially prevalent among workers who are in permanent contact with people (Maslach et al., 2001).

Structural empowerment focuses on conditions in the work environment, whereby power, decision-making, and formal control over resources are shared (Kanter, 1977). According to an individual with access to information, resources, support, and opportunities is in a better position to learn and develop. Access to these structures of empowerment facilitates better performance in the workplace as well as in interpersonal relationships which are assisted by effective and efficient communication. Kanter argues that an empowered environment could improve the working conditions that lead to burnout in workers. Enhanced SE with good support and resources from the organization will facilitate an appropriate information flow and will promote development opportunities for the employees. These organizational characteristics constitute a genuine antidote to burnout. High levels of SE contribute to protecting the subjects from suffering burnout (Leiter et al., 2012).

Psychological empowerment refers to the beliefs a worker has regarding their competence, autonomy, and the outcomes that their work provides to the organization. The focus of PE is on the state or set of conditions that allow for employees to believe that they have control over their work. The four aspects of PE are meaning, impact, competence, and autonomy. Meaning involves consistency between the requirements of the post and the beliefs, values, and behaviors of the worker. Impact is associated with the sense of being able to contribute to significant outcomes for the organization. Competence is linked to confidence in having the personal skills to perform the particular job. Autonomy means that the individual has the initiative to choose and organize their actions, this being an important mechanism for reducing tension and stress (Spreitzer, 1997). PE determines the attitudes and individual behaviours which contribute to lower levels of burnout at work

Psychological empowerment is a logical outcome of SE. In turn, PE has beneficial effects on burnout of employees, acting as a protective factor in reducing the effects of stressors of the work environment on burnout (Laschinger, Finegan, Shamian, & Wilk, 2004; Laschinger et al., 2003. PE mediates the relationship between the dimensions of SE and two dimensions of burnout (emotional exhaustion and professional efficacy/personal accomplishment), helping to reduce chronic stressors in the workplace.

Managers' hardiness effect

Hardiness is a personality composite of beliefs about self and world involving the importance of a sense of commitment, control, and challenge (Kobasa, Maddi, & Kahn, 1982). Commitment implies that the subject takes an active part in all their daily tasks, being able to identify themselves with their own work. This commitment involves the personal recognition of their own objectives and decision-making while maintaining their values. Control refers to the ability to intervene when dealing with the development of events. The person feels responsible for what happens. Challenge is seen as the opportunity that situations of uncertainty provide to improve their own abilities. These challenges of change create a flexible and tolerant personality who experiences uncertainty as a step forward, avoiding living in a permanent state of stress. According to Hystad, Eid, and Brevik (2011), these three factors are essential in order to respond successfully to the high demands which workers are subject to in their work.

According to Menezes (2012), higher SE generates new action possibilities but also some stress (higher perceived demands, some ambiguity pertaining to job scope/action possibilities). People with high hardiness levels are better equipped to face this stress and to take better advantage of the conditions of SE (support, resources, information, development opportunities) in comparison with people with less hardiness. Subjects with a hardy personality possess a cognitive, attitudinal, and behavioral flexibility, as well as a high tolerance of ambiguity. They are people who know how to cope with unforeseen events in the workplace and who try to seek alternative solutions to resolve the problems about which they can do something; they understand stressful stimuli at work as opportunities to grow and as a means of learning when faced with current and future events (Kobasa, 1979); they believe that everything can be learned from and that they have the personal resources to do so.

In consequence, it is expected that individuals with high hardiness (who are compromised, have a confidence in their capacity to do their job, and perceive difficulties as opportunities to grow) will take a better advantage of work environment conditions and will show a higher PE than individuals with a low level of hardiness. The individuals with access to information who, moreover, show high control and compromise will see their job as a personally significant one because they understand how their role at work fits in with the goals and strategies of the organization. Besides, the individuals with access to resources who, moreover, show high control will feel responsible for what happens and will believe their actions to have a significant impact on work unit and/or organization. Furthermore, the more compromised they are, individuals will get further support from their workmates and the organization, which will in turn contribute for them to experience a further sense of autonomy because, as compromised and widely accepted members of the organization, they will have more options to choose and organize their actions (Menezes 2012),

People with high levels of hardiness believe that they are able to efficaciously handle the information, resources, and supports which are available at their work environment and envision those as opportunities to further learn and improve, which also develops their sense of helpfulness in the attainment of significant results for the organization. The basic proposition is that the individuals with hardiness view their work environment as a space of proving opportunities rather than one full of constraints on individual behavior, thus enhancing their positive outcomes and contributing to advance their PE. In short, the hardy personality interacts with SE and contributes to a greater psychological empowerment (Menezes 2012),

Based on coping literature we would expect that individuals with low hardiness would be those who could benefit themselves further from betterment in the psychological empowerment. However, it requires a minimum level of hardiness in order to have confidence in one's own abilities and to be able to experience psychological empowerment the position of the middle manager requires people with the ability to regulate their own demands and those of others in the interpersonal transaction processes that occur in the workplace. The hardy personality shows evidence of personal ability in exercising control over their own thoughts and the actions they carry out. In complex situations, they react in trying to solve them in the most constructive way possible. Middle managers with hardiness defy adversity, become involved, and approach change and stressful situations naturally (Eschleman, Bowling, & Alarcon, 2010). The hardy personality can help middle managers improve their responses to the conditions of the working environment and thus increase their PE and reduce their levels of burnout.

As new days organizations restructure and re- forming in the name of efficiency and effectiveness, trust in management has become an increasingly important element in determining organizational climate, employee performance and commitment to the organization. Employees who have survived downsizing are understandably wary about the future direction of the organization and their roles within it. Organizational trust as the extent to which one is willing to ascribe good intentions to, and have confidence in the words and actions of other people. Trust has a significant impact on important organizational factors such as group cohesion (Podsakoff, MacKenzie & Bommer, 1996), perceived fairness of decisions, organizational citizen-ship behavior, job satisfaction, and organizational effectiveness. Mistrust results when information is withheld, resources are allocated inconsistently, and when employees have no support from management.

Without trust, people cannot or will not work together except under conditions of stringent control (Whitney, 1994). Ironically, at a time when trust is most needed for successful organizational transformation, the changes resulting from restructuring have diminished trust within the work setting. This state of affairs has serious implications for organizational performance (Ouchi, 1981). Nurses, the largest group of health-care providers in hospitals, have been particularly hard hit by recent downsizing. It is quite possible that their mistrust of the system could potentially threaten the quality of patient care.

Kanter (1977, 1993) maintains that work environments that provide access to information, resources, support and the opportunity to learn and develop are empowering and enable employees to accomplish their work. As a result, employees are more satisfied with their work and sense that management can be trusted to do whatever is necessary to assure that high quality outcomes are achievable. According to Kanter, employees in environments such as these are more committed to the organization and more likely to engage in positive organizational activities. Kanter's theory provides an explanatory framework for investigating the role of trust in empowering nurses to function effectively in today's dramatically restructured health-care settings. The purpose of this study was to test a model linking staff nurses' workplace empowerment, organizational trust, job satisfaction, and organizational commitment.

Organizational Empowerment

Kanter (1977, 1993) suggested that people react rationally to the situations in which they find themselves. When situations are structured in such a way that employees feel empowered, the organization is likely to benefit both in terms of the attitudes of employees and the organization's effectiveness. In fact, Kanter argues that the impact of organizational structures on organizational behavior is far greater than the impact of employee personality predispositions. The organizational structures that Kanter believes particularly important to the growth of empowerment are: having access to information, receiving support, having access to resources necessary to do the job, and having the opportunity to learn and grow . Access to these empowering structures is facilitated by formal job characteristics. That is, jobs which are visible and central to the organiza-tion's goals and which allow the employee flexibility enhance empowerment . In addition, informal job characteristics such as alliances with superiors, peers and subordinates within the organization further influence empowerment. According to Kanter, the mandate of management is to create conditions for work effectiveness by ensuring employees have access to the information, support, and resources necessary to accomplish work and are provided ongoing opportunities for employee development. Having access to these structures results in increased levels of organizational commitment, feelings of autonomy, and self-efficacy. Consequently, employees are more productive and effective in meeting organizational goals. Relationships among constructs in Kanter's theory are presented in figure (1).

Figure (1): Final standardized model tested with productivity outcome variable (Kanter's theory).

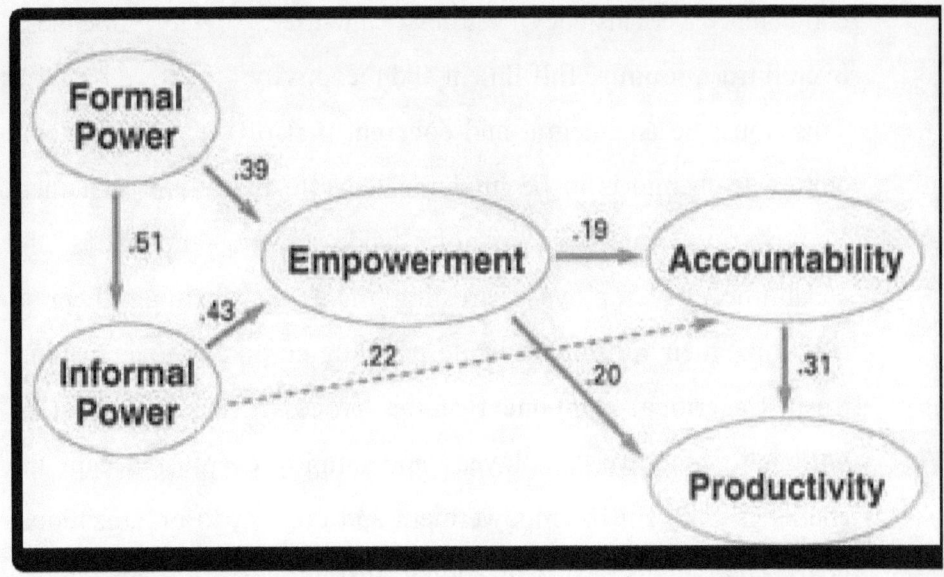

Institutional Trust

Trust is increasingly important to organizational relationships, particularly in light of dramatic organizational changes designed to flatten organizational structures and place more decisional control in the hands of front-line employees (Hart, Capps, Cangemi & Caillouet, 1996). According to Kanter, trust evolves from a mutual understanding based on shared values and is essential for employee loyalty and commitment. Organizational trust is defined by Gilbert (1995) as the belief that an employer will be straightforward and follow through on commitments. Trust refers to employee faith in organizational leaders and the belief that ultimately organizational actions will prove beneficial for employees. Mishra and Morrissey (1990) argue that open communication, sharing of critical information, sharing of perceptions and feelings, and greater worker involvement in decisions facilitate trust in organizations. Identified conditions of trust: discreetness, availability, competence, consistency, fairness, integrity, loyalty, openness, overall trust, promise fulfillment and receptivity.

Trust must be an integral and coherent part of the organizational culture if change is to be implemented effectively and sustained. Empowering employees involves understanding the needs and capabilities of employees, trusting them, and helping them to maximize their fulfillment while pursuing corporate goals. Mutual trust is a critical component of this process. Managers must be willing to empower employees and employees must accept the challenge inherent in empowerment and commit to organizational goals. High levels of organizational trust are needed to accomplish change, yet paradoxically, the change itself may destroy trust and threaten organizational effectiveness. Research on organizational downsizing has shown that decreased levels of trust are associated with decreased communication and increased conflict (Mishra & Spreitzer, 1998). As hospitals continue to downsize, employee trust and morale are eroded as workloads increase and job insecurity escalates. In such low trust organizations, behaviors such as high absenteeism, prolonged breaks, limited learning, low accountability, reactionary thinking, and low creativity are predictably common (Cangemi, Rice & Kowalski, 1989).

Kramer, Brewer, and Hanna (1996) maintain that employees in low power/low status positions depend on others for a variety of critical organizational resources and that uncertainty limits access to information needed to make judgments about trustworthiness. Similarly, Daley (1991) claims that vulnerability and uncertainty are central to the issue of trust and that violations of trust take on greater significance for those in relatively low power/control positions, such as hospital staff nurses . According to Tyler and Degoey (1996), managers play a crucial role in the development of trust since they control the flow of information by either sharing or not sharing key information. The degree of trust within an organization depends on managerial philosophy, organizational actions and structures, and employees' expectations of reciprocity. Gilbert (1995) found a strong positive relationship between organizational trust and the nature and extent of organizational communication. They suggest that formal, but even more importantly informal, access to organizational communication channels enhances organizational trust. Podsakoff et al. (1996) found that transformational leadership behaviors accounted for 28% of the variance in trust in management.

The impact of trust on organizational outcomes has been reported in the organizational literature. Podsakoff et al. (1996) found organizational trust to be significantly related to job satisfaction, organizational commitment, role clarity, and in-role performance. Andrews (1994) linked employee empowerment in a large manufacturing firm to an atmosphere of mutual trust. Founded that 90% of managers surveyed in their study felt that trust starts at the top of an organization and trickles down. Organizational effectiveness was perceived to depend on the level of organizational trust. Trust was associated with effective decision-making as a result of sharing ideas, information, and feelings, organizational credibility, and increased productivity. Organizational ineffectiveness was attributed to employee distrust of management by 79.4% of those surveyed. Concluded that organizational empowerment and trust have a significant impact on job design, control mechanisms, extent and effectiveness of communication, relationships with other units, and degree of innovation, job satisfaction, commitment, organizational citizenship behaviors, goal sharing and crisis management.

There is little empirical research in the nursing literature relating to organizational trust. In a survey of staff nurses in a small U .S. hospital, Kramer and Schmalenberg (1993) concluded that trust was the best predictor of feelings of autonomy and empowerment. In another nursing study, McDaniel and Stumpf (1993) found that nurses were more empowered in health-care organizations where information is shared and trust levels are high. Emphasized the importance of leadership behavior in developing and maintaining trust levels in nursing work settings. Finally, anecdotal accounts from nurses reflecting distrust of current management systems are prevalent in today's turbulent nursing work environments.

Leaders and Organizational Commitment

Organizational commitment consists of employees' attachments to their organization (Buchanan, 1974; Porter, Steers. there are three types of organizational commitment. Affective commitment is an individual's emotional attachment, identification with, and involvement in a particular organization. Employees with strong affective commitment work in the organization because "they want to". Continuance commitment reflects an employee's awareness of the costs associated with leaving an organization. Individuals with high continuance commitment believe the benefits of staying with an organization outweigh the consequences of leaving and stay with the organization because "they need to." This type of commitment is likely to be prevalent in today's downsized work environments. Normative commitment reflects an individual's sense of obligation for remaining in the organization.

Affective commitment has many positive consequences for the organization. It has been found to be positively related to job satisfaction, job involvement, job performance and organizational citizenship behavior. Employees with strong affective commitment contribute more to the accomplishment of organizational goals. They are also less likely to leave the organization. Although employees with high continuance commitment are also less likely to leave the organization, this lower turnover occurs at the expense of employee engagement, job satisfaction, and self-esteem (Sethi, Meinert, King & Sethi, 1996) . Employees with high continuance commitment may be motivated to do the minimal amount of work required to maintain their jobs. Thus, this type of commitment may be counterproductive to the effective accomplishment of organizational goals and objectives (Meyer & Allen, 1997; Sethi et al., 1996). This claim was corroborated by McCloskey and McCain (1987) who found that nurses who stayed in their positions because of job scarcity reported higher levels of absenteeism and were increasingly more likely to demonstrate poor work performance.

Organizational commitment is of particular importance to health-care organizations. Employees in these turbulent environments are struggling to maintain high quality patient care with fewer resources. The empirical evidence suggests that employees with high affective commitment are more likely to rise to the challenges imposed by restructuring while employees with high continuance commitment may simply do the "minimum." Moreover, Glisson and Durick (1988) found that individuals displaying higher levels of affective commitment were more resistant to job strain and burnout suggesting that affective commitment may help employees withstand the negative effects of downsizing. Clearly, it is important for health-care organizations to promote those factors that encourage affective commitment and to reduce those that encourage continuance commitment (Menezes 2012).

Work experiences are the strongest predictors of affective commitment. In a meta-analysis of organizational com-mitment research, Mathieu and Zajac (1990) found that job scope, challenge, and high levels of work autonomy were consistently related to commitment. Bateman and Strasser (1984) found higher levels of affective commitment when exceptional job performance was recognized and rewarded. Opportunity for advancement, perceptions of fairness regarding the distribution of rewards, and workplace autonomy have been found to positively influence staff nurses' organizational commitment (Curry, Wakefield, Price, Mueller & McCloskey, 1985). These findings are consistent with Kanter's (1977) contention that structural factors of the work environment contribute to an individual's affective commitment to the organization.

That said, Meyer, Irving and Allen (1998) note that although such positive experiences contribute to affective commitment, these "same experiences were found to be unrelated, or negatively related, to continuance commitment" (p. 33). Indeed, a different set of factors is likely to lead to continuance commitment. Meyer, Bobocel and Allen (1991) argued that continuance commitment was related to the perceived availability of alternatives and to the investments that would be lost if the employee left the organization. These arguments suggest that empowerment is more likely to influence affective commitment, and that its relationship with continuance commitment should be minimal at best.

Elderly care in Sweden has been subjected to many changes in recent years, including reduced funding and staffing shortages. This leaves managers to cope with multiple demands that sometimes exceed the available resources. According to Kanter's theory of structural empowerment, structural conditions are essential to employees' well-being and organizational effectiveness. However, first-line managers (FLMs) have been found to perceive lower structural empowerment when com-pared with their superiors. Most of the previous research has been based on female FLMs in health care. Therefore, it is of interest to explore FLMs' work situation and access to sup-portive structural conditions in elderly care from the perspective of male managers (Swedish National Board of Health & Welfare 2011).

The role as manager involves dealing with, for example, finances, business strategies, organizational operations, human resources and information technology. Qualitative studies (Shirey et al. 2008, McCallin & Frankson 2010) conducted in hospital settings with mostly female participants have described the managerial role as over-whelming, with unrealistic expectations that tend to exceed the available resources. Their role was often unclear, involving difficulties in task reallocation (Reay et al. 2003) that resulted in feelings of work-related stress. Managerial work has been found to include conflicts concerning distribution and fragmentation of time, leading to a difficult balance between work and private life and between work assignments (Tengelin et al. 2011). Stressful work has in turn increased the risk of illnesses, e.g. severe burn-out, emotional and physical fatigue, insomnia, impaired psychological health and work–private life imbalance (Macphee et al. 2012). However, positive work environments have also been reported, as well as moderate to high job satisfaction (Abdelrazek et al. 2010) and positive relationships between organizational support, empowerment, degree of control and job satisfaction (Lee & Cummings 2008).

Studies comparing male and female employees, in different areas, have found that work conditions affect men's health, and in particular that low stress at the workplace is the most important predictor of good health among men (Hakansson & Ahlborg 2010), male white-collar employees who worked a great deal of overtime had a higher risk of displaying health symptoms than did their female colleagues. Compared with women white-collar employees, male white-collar employees have a statistically significantly higher risk factor for sickness absence due to conflicts between the demands of their work and those of their private life. Regarding male managers, they have been reported to be less effective than female managers. However, a prospective study found no differences in turnover rates between male and female managers

(Skagert et al. 2011). Earlier research also showed that being a male manager can be more favourable, as men in managerial positions show lower stress levels at work (Lundberg & Frankenhaeuser 2018) than do their female counterparts.

The concept of empowerment has been widely used but never adequately denned. Since the ideology has been adopted to promote the rights of ethnic and sexual minorities, for training and education programs as well as in organizational development programs, and by the feminist movement. Recently the concept has also appeared more frequently in the nursing literature. There have been some concept analyses (Gibson 2017) and studies exploring the problems associated with empowerment and the concept of power itself (Gilbert 1995). However, these analyses have failed to provide a coherent view on the content of the concept.

Two indices. The earliest papers, from the 1980s, consist of no more than 25 articles, which deal with empowerment primarily from the point of view of the development of the nursing profession. The number of articles on empowerment has sharply increased during the 1990s. Over half of the latter discuss empowerment in the context of the patient nurse relationship (e.g. in the treatment of psychiatric and HIV patients), preventive nursing and health education. The concern in this paper, however, is with the articles dealing with organizational and professional development. These represent about one-third of the papers produced by the literature search. A total of 46 articles were concerned specially with questions of professional growth and development in nursing.

The majority of literature were divided the basis of theoretical orientation into three categories and led of application. The first category leans heavily on critical social theory and emancipatory theory, e.g. feminism. Of the 46 articles in our review, 14 are included in this category. Here empowerment is associated with improving the living conditions of oppressed groups such as racial minorities, women and health care patients. The second category of empowerment (18 articles) consists of organization theories, where empowerment is associated with the delegation of power and the subject's opportunity to take action. The third category is concerned with social psychological theory, based on the individual's development. This theory suggests that empowerment originates within the individual and is concerned with the individual's resection within the environment. There are 14 articles in the third category. Those based on critical social theory stand clearly apart as an independent category. The philosophy of the articles based on leadership and social psychological theories is partly the same. One possible reason why articles in the second category outnumber those in the first lies in the current popularity of empowerment as a management tool in the business world.

It is clear from the growing number of papers dealing with empowerment that the concept has been more and more widely adopted in nursing research. Empowerment is a useful umbrella concept to describe the elements of professional growth and development in the nursing profession. In some studies, however, definitions of empowerment have been so broad and sweeping that it has become almost synonymous with the concept of nursing care. This is bound to undermine the value and utility of the concept. It is crucially important that the concept is properly defined. The classification proposed above between critical social theory, organization theory and social psychological theory may be helpful when the concept is used as a frame of reference for research and development projects (Menezes 2012),

Empowerment is also a useful conceptual innovation for organization theories. Here empowerment provides well-being at both the individual and organizational level, which ultimately reinforces staff self-images and cooperation networks. It may be assumed that management influenced by this idea of empowerment will serve to strengthen staff nurses' professional self-esteem, which in turn will contribute to professional growth and development. Staff cannot, however, be empowered merely by delegation, by transferring tasks downwards in the organizational hierarch. (Menezes 2012).

As the authoritarian type of leadership assumes less arbitrary forms, the individual's personal qualities and ways of acting assume increasing importance. Voluntary spontaneous activity as well as a willingness and ability for self-improvement are expected of individuals. Nursing is unique as an occupation in being grounded almost entirely on human relations, and in this sense it differs completely from industrial production. It would be interesting to identify which variables inherent in the nurses themselves and in the nursing environment serve to enhance nurses' professional development and inner feelings of strength. Given its emphasis on the individual and environmental factors, social psychological theory provides a suitable framework for such studies. The information obtained would be extremely useful for the development of both organizations, staff and nursing care. The factors involved in empowerment may also open up new perspectives for the development of quality nursing care.

The nursing profession is facing a potential workforce shortage as many nurses near retirement and recruitment, and retention of nursing staff has become a priority in health-care organizations. Furthermore, numerous studies have identified factors influencing staff nurse recruitment and retention. However, less attention has been given to recruitment and retention of nurse managers. In Canada, nurse manager positions have changed dramatically following extensive restructuring in the 1990s that resulted in fewer nurse man-agers with increased responsibilities, highly demanding jobs, and stressful working conditions (Canadian

Laschinger et al. (2008) found that nurses in management roles in hospitals averaged 48 years in age, with little difference among levels of management, suggesting that the nursing leadership workforce may experience a sudden and severe shortage within the next decade. Furthermore few staff nurses are expressing interest in moving into these management positions (Rudan 2002, Wong et al. 2011), citing unreasonable workloads and unrealistic job expectations of these roles.

The impending retirement of a large proportion of the current cohort of nurse managers and the perceived unattractiveness of management jobs to staff nurses present a challenge to sustaining strong leadership in future nursing work environments and highlight a need to identify factors that promote retention of current nurse managers. Empowering leadership has played an important role in promoting retention of staff nurses. Laschinger et al. 2009) but only a few studies have linked leadership to nurse manager work-life quality. There is limited research that explains the mechanisms by which leadership influences nurse manager's intentions to leave or remain in their jobs. Therefore, the purpose of this study is to examine the influence of senior nurse leadership practices on nurse managers experiences of empowerment and support in the workplace and, ultimately, on their perceptions of patient care quality and turnover intentions

.

Leadership model

Kouzes and Posner s model of leadership is based on five fundamental leadership practices that enable leaders to accomplish extraordinary things in organizations. The current study defines leadership as the art of mobilizing others to want to struggle for shared aspirations. This leadership model consists of five best leadership practices: challenging the process, inspiring a shared vision, modeling the way, enabling others to act and encouraging the heart. Challenging the process is exemplified by the core behaviors of seeking out opportunities for change, questioning the status quo, taking risks to improve the process, demonstrating flexibility, creative thinking and taking initiative (Kouzes & Posner 1987). Inspiring a shared vision is a leadership practice reflected in the core behaviors of positive communication, active listening, interpersonal competence, commitment to a vision, and sharing of a common purpose. Modeling the way reflects the core behaviors of setting an example, clarifying values, sustaining commitment, making a plan and achieving small wins. Enabling others to act reflects the core behaviors of collaboration, building trusting relationships, sharing information and resources, developing leadership of self and providing visible support. Finally, encouraging the heart is a leadership practice that involves recognizing contributions, building confidence, providing feedback, celebrating accomplishments and creating social support (Kouzes & Posner 1987).

Structural empowerment

According to Kanter (1977, 1993), structural factors within the work environment have a greater impact on employee work attitudes and behavior than personal predispositions or socialization experiences. The organizational empowerment structures described by Kanter are: access to information, support, resources needed to do the job and opportunities to learn and grow. Access to information means having knowledge of organizational decisions, policies and goals, as well as data, technical knowledge and expertise that enable one to be effective within the broader context of the organization. Information provides employees with a sense of purpose and meaning, while enhancing their ability to make informed decisions and contribute to organizational goals.

Access to support includes informal and formal feedback and guidance received from superiors, peers and subordinates. Access to re-sources refers to individual's capacity to access the materials, money, supplies, time and equipment required to accomplish organizational goals. Opportunities for learning and growth include access to challenging work, rewards and professional development opportunities. Formal and informal job characteristics facilitate employee access to these empowerment structures. Formal power is enhanced by flexible jobs central to the organization s goals that allow employees to implement creativity and discretionary decision-making. Informal power comes from developing effective relationships and communication channels both within and outside the organization (Kanter 1977, 1993).

Numerous studies have linked Kanter s concept of structural empowerment to employee attitudes and behaviors. These include job satisfaction, job control and organizational justice, as well as occupational health variables such as job strain (Laschinger et al. 2001) and burnout (Laschinger et al. 2003). A few studies of nurse manager empowerment have shown that structural empowerment increases as one climbs up the organizational hierarchy. Accordingly, middle managers reported significantly higher levels of empowerment than first-line managers, who reported higher levels of empowerment than staff nurses. Empowered man-agers can empower others and are more likely to do so than those who do not feel empowered. When nurses perceive their managers to be confident and have influence in the organization, they are more likely to feel empowered themselves (Laschinger & Shamian 2016)

Managers' Perceptions of organizational support

Perceived organizational support refers to employees' general beliefs about the degree to which their organization for which they work values their contributions and cares about their wellbeing. Perceived organizational support is a fundamental condition of employee satisfaction and attitudes towards their work. The organization also reaps rewards of high POS, as it results in greater organizational commitment and increased efforts on the part of the employees to fulfill organizational goals. A review by Rhoades and Eisenberger (2002) high-lighted a strong relationship between POS and several important occupational outcomes: affective organizational commitment, overall job satisfaction, positive affect and fewer symptoms of job strain, such as emotional exhaustion, fatigue and anxiety. Higher levels of POS were associated with enhanced performance of job duties, going above and beyond role expectations, and risk reduction behaviors. Moreover, turnover inten-tions, absenteeism and lateness were all significantly and negatively related to POS (Rhoades & Eisenberger 2002).

Perceived organizational support has been examined in several studies of nurse managers. Tansky and Cohen (2001) found POS to be significantly related to nurse manager s organizational commitment and Laschinger et al. (2006) showed that POS was positively related to nurse manager s perceived rewards for ef-fort, respect, job security and quality of care ratings. Patrick and Laschinger (2006) linked structural empowerment to POS ($r = 0.65$) and both these variables positively predicted nurse manager role satisfaction. Thus, evidence supports links among structural empowerment, POS and manager outcomes such as organizational commitment, perceived care quality and role satisfaction.

Managers' empowerment and patients' Quality of care

Perceptions of quality of patient care have been related to structural empowerment and organizational support in the nursing profession. In a large international study Aiken et al. (2002) showed that nurses were twice as likely to rate the quality of care as fair or poor when organizational support for their work was rated lowest. Empowering work environments have been associated with better quality of patient care (Laschinger 2008, Purdy et al. 2010). Purdy et al. (2010) found that structural empowerment at the team level had a significant indirect effect (b = 0.39, P < 0.001) on nurse-assessed quality through group processes, and Laschinger (2008) found a significant effect of structurally empowering working conditions on nurse-assessed care quality (b = 0.29, P < 0.05) in a study of Ontario acute-care nurses. Perceived organizational support has been significantly but weakly (r = 0.19) positively associated with nurse manager-assessed quality of care (Laschinger et al. 2006). Finally, qualitative studies have found that the ability to ensure quality of care was an important factor in retaining nurse managers (Menezes 2012).

Healthcare organizations and leadership have changed in recent decades. Reductions in financial resources, advances in techniques, and a rapid increase in and new forms of information have contributed to these changes. Organizational and professional changes have had an impact on the role of the nurse manager. Head nurses are expected not only to manage the care of patients, but also to lead their departments professionally and administratively. Leadership is a key factor in creating empowering conditions in the workplace (Laschinger et al., 2009). Furthermore, empowerment is presented as the way in which organizations can move from the static and rule-bound past to the dynamic and flexible future (Procter et al., 2015).

Empowerment has become a widely-used concept in nursing, but due to its ambiguity, it is difficult to define, as it assumes different forms in different contexts (Bradbury-Jones et al., 2008). In addition, empowerment as a goal is described in the literature as power, control, ability, competence, self-efficacy, autonomy, knowledge, development, self-determination, and strengthening of the position of one's own group in society (Arneson & Ekberg, 2006). Work-related empowerment can also be conceptualized as three unique but interdependent concepts: verbal, behavioral, and outcome empowerment. Several concept analyses of empowerment in nursing have been published and empowerment has been conceptualized in different ways in the literature. The three theoretical approaches most often used in the published literature are: critical social theories (Gibson, 2017).

Social psychological theories Those who link empowerment to critical social theory believe that it can only be understood in relation to the history and structure in which nurses' find themselves (Fulton, 1997). In the context of nursing, the theory emphasizes the concept of empowerment in different social situations, such as among nurses, between nurses and patients, and namely other healthcare professionals (Kuok-kanen & Leino-Kilpi, 2000). From a psychological perspective, empowerment is viewed as the perception or attitudes of individuals towards their work and their role in the organization. It is noted that the structural factors within the work environment have a greater impact on the work attitudes and behaviors of employee than personal pre-dispositions or socialization experiences. The organizational empowerment structures described by Kanter are: access to information, support, resources needed to do the job, and opportunities to learn and grow (Laschinger et al., 2007).

For managers, the term "to empower" has traditionally meant "to authorize", "to give authority to others", or "to invest power in another person". When the nurse manager uses the term "empower", it automatically denotes a kind of unequal relationship between a superior and a subordinate. In healthcare settings, an unequal power base exists among administrators, physicians, and nurses as a result of competing goals of administration and the coexistence of multiple lines of authority. The rigidity of hierarchical rule-bound structures specifically has been blamed for nurses' inability to control the content of their practice sufficiently (Kerfoot, 1994).

Managers are ideally positioned to create the structural conditions for work effectiveness. In addition, Laschinger et al. (2007) cite Kanter (1977) described the organizational empowerment model where structural factors, such as access to information, support, resources, and opportunity in the work setting, are posited to have a major influence on the employees' ability to accomplish their work. Furthermore, workplace empowerment is a management strategy that has been proven to be successful in creating a positive work environment in organizations (Laschinger et al., 2009). In organizational settings, empowerment creates and sustains a work environment that facilitates the employee's choice to invest in personal actions and behaviors, resulting in a positive contribution to the organization's mission (Marquis & Huston, 2000). Nurses and nurse managers who experience strong empowerment have qualities that form a strong sense of self-esteem, successful professional performance, and progress in their work. Higher unit-level structural empowerment has also been found to be associated with lower emotional exhaustion (Laschinger et al., 2011).

Some empirical studies on nurse managers' empowerment have been published, but no systematic reviews of the topic were found. However, a systematic review of front-line man-agers' job satisfaction has been published, in which an association between job satisfaction and empowerment was noted. The findings of the above-mentioned review indicated the existence of a significant, positive relationship between empowerment, span of control, organizational support, and job satisfaction (Lee & Cummings, 2008). Further improvement is needed in the area of empowerment at the management level, because those nurse managers who feel empowered in their nursing leadership roles do not only have access to more power themselves, but are also able to support the mechanisms that would offer employees an opportunity to empower themselves, and accordingly, this could make an important contribution to the healthcare setting in hospitals where the nurse managers work.

Nurse Managers' empowerment factors

The empowerment of nurse managers has been observed to be related to different variables, such as demographic characteristics (unit speciality, education) (Laschinger, 2006), experience of stress, need for further education in leadership and development, knowledge and skills in handling the job, work motivation, and the importance of work independency (Suominen et al., 2005). Likewise, empowerment correlated positively with job satisfaction (Laschinger et al., 2004; Suominen et al., 2005), per-ceived organizational support, role satisfaction (Patrick & Laschinger, 2006).

Despite of workplace empowerment researches revealed a highly significant link between empowerment and positive work behaviors and attitudes. Researches demonstrating the essential relationship between structural empowerment and psychological empowerment will provide direction for future interventions aimed at the development of a strong and effective health care sector . Methods Published research articles examining structural empowerment and psychological empowerment for nurses were selected from computerized databases and selected websites. Data extraction and methodological quality assessment were completed for the included research articles .Results Ten papers representing six studies reveal significant associations between structural empowerment and psychological empowerment for RNs . Implications for nursing management Creation of an environment that provides structural empowerment is an important organizational strategy that contributes to RNs psychological empowerment and ultimately leads to positive work behaviors and attitudes. Critical structural components of an empowered workplace can contribute to a healthy, productive and innovative RN workforce with increased job satisfaction and retention (Suominen 2005).

The Canadian health care system is under constant stress from numerous environmental factors: an ageing workforce, an increase in morbidity associated with ageing population, rapidly advancing technology and exponential advances in knowledge (Canadian Institute for Health Information 2002). Recent studies report that registered nurses (RNs) are working with more complex patients, have fewer available resources, and have reduced opportunities to take time off for education, training and placements as a result of health restructuring (Canadian Institute for Health Information 2002). The Canadian Institute for Health Information also reports that health care workers, including nurses, are more likely to have stress or job strain-related absences from work than workers in other sectors. At the same time, Health Canada (2006) reports that one-third of Canadian RNs are aged 50 years or older, and many are considering early retirement. Registered nurse recruitment and retention issues have become a major concern for health care leaders.

Researches on structural empowerment in health care settings (Laschinger 2008a) indicates that changes in workplace structure can support healthier employees, reduce stress and increase employee commitment to organizational goals, culminating in improved organizational outcomes that include improved patient care. According to Kanter (1977, 1993), there are two systemic sources of power in organizations: formal power associated with jobs that have high visibility, are essential to the organization and require independent decision-making; and informal power – derived from relationships or alliances with superiors, peers and subordinates (Miller et al. 2000). Formal and informal power facilitates access to job-related empowerment structures of :support and feedback and guidance received from superiors, peers and subordinates information the data, technical knowledge and expertise required to function effectively in ones' position resources the time, materials, money, supplies and equipment necessary to accomplish organizational goals and opportunity autonomy, growth, a sense of challenge and the chance to learn and grow (Havens & Laschinger 1997).

The University of Western Ontario Workplace Empowerment research program is a program of research, based on Kanter s (1977) original theory that emphasizes that staffs need increased access to opportunity, information, resources, support, formal power and informal power if they are to be empowered. These six components of structural empowerment have been identified through extensive research as separate and distinct sources of organizational power. Research also reveals that the specific behaviors and attitudes of job satisfaction, commitment, trust and low burnout are influenced by all six components of structural empowerment (Beaulieu et al 2004).

Expanded workplace empowerment model where Spreitzer (1995) model of psychological empowerment provided an intervening role between structural empowerment and job satisfaction. Psychological empowerment is an essential component of workplace empowerment, representing intrinsic task motivation, or employee rewards that are inherent to empowering work conditions (Laschinger et al. 2009). Components of this multi-faceted construct of psycho-logical empowerment include: meaning – a fit between job requirements and beliefs, or the value of a work objective, compared with an individual s own ideals or standards; competence an individual s confidence or belief in their abilities to perform activities with proficiency; self-determination – sense of choice or control over one's work/autonomy and in the commencement and maintenance of work activities in the workplace; and finally, impact – the sense of being able to influence important outcomes at work. These authors stressed that the four dimensions reveal an orientation towards work reflecting the individual s desire and ability to influence his or her job and workplace (Thomas & Velthouse 1990).

Identifying and understanding the relationship between structural empowerment and psychological empowerment will assist health care leaders to counteract the impact of environmental stress on the health care sys-tem and to improve the recruitment and retention of nurse managers. Structural empowerment, or the individual s awareness of empowering workplace surroundings (Laschinger et al. 2009), has been demonstrated to have significant measurable impact on health care personnel when psychological empowerment, or the psychological state that employees must experience for empowerment interventions to be successful (Laschinger et al. 2001), is also present. Research demonstrating this essential relationship between structural empowerment and psychological empowerment will provide direction for future interventions aimed at the development and maintenance of a strong and effective health care sector. A health care sector that supports healthier employees reduces stress and increases employee commitment will culminate in improved organizational outcomes, including improved patient care (Laschinger 2008). Dense regarding: the relationship between components of structural empowerment and overall psychological empowerment; and, the relationship between overall structural empowerment and components of psychological empowerment provided valuable insights into the Ontario nurses workplace. Different methods path analysis, structural equation modeling (SEM), multilevel structural equation modeling, hierarchic regression and product variable approach regression.

Related Studies

A quantitative component of a mixed methods study using survey principles. Methods. The conditions of work effectiveness questionnaire was distributed to the total population (n = 517) of nurse leaders in an NHS Foundation Trust in England. Nurse leader groups comprised unit leaders (sisters, matrons) and senior staff nurses. Results Overall, the unit response rate was 441% (n = 228). Levels of total and global empowerment were moderate and moderate to high respectively. Groups did not differ significantly on these parameters or on five elements of total empowerment, but significantly higher scores were found for unit leaders' access to information. Significantly higher scores were found for senior staff nurses on selected aspects of informal power and access to resources, but scores were significantly lower than unit leaders for components of support Conclusions moderately empowered population of nurse leaders differed in relation to access to information, aspects of support, resources and informal power, reflecting differences in roles, spheres of responsibility, hierarchical position and the constraints on empowerment imposed on unit leaders by financial and resource pressures. (Laschinger 2006, Armellino et al. 2010).

A study of nursing job satisfaction has focused on both outcomes and antecedents of job satisfaction/dissatisfaction. In a meta-analysis of 48 studies, Blegen (1993) identified 13 predictors of nursing satisfaction. These included personal attribute variables or personality traits such as age, education, years of experience and locus of control, and organizational variables such as super-visor communication, commitment, stress, autonomy, recognition, reutilization, peer communication, fairness and professionalism. Organizational variables were more strongly related to job satisfaction, with correlations ranging from highs of -0 .61 and 0 .53 for stress and commitment respectively to a low of 0.36 for communication with peers. Personal attribute variables had considerably the association between job satisfaction and turnover among nurses is well supported in the literature. Irvine and Evans (1995) found that job satisfaction was strongly related to intention to leave (r = -0 .53) in a meta-analysis of studies Opportunity to move into another job was a modifier of job satisfaction/turnover intention relationship. Few studies have linked nurse work satisfaction to client outcomes. Weissman and Nathanson (1985) found that job satisfaction of 344 registered nurses employed as primary providers in 77 Maryland family planning clinics predicted patient satisfaction which, in turn, predicted patient compliance with prescribed contraceptives. However, could find no significant relationships between nurses' job satisfaction, quality of nursing care, and patient satisfaction in an urban Midwestern teaching hospital. Methodological constraints make it difficult to establish the nature and direction of relationships between nurses' job satisfaction and client outcome (Menezes 2012).

Study a predictive, non-experimental design was used to test Kanter's work empowerment theory in a random sample of 412 Canadian staff nurses. Empowered individuals reported higher affective commitment and work satisfaction. Moreover, empowered employees experienced greater organizational trust, which in turn influenced these job attitudes. Since research has shown that affective commitment is related to productivity, our results suggest that fostering environments that enhance perceptions of empowerment will have positive effects on employees and ultimately, enhance organizational effectiveness.

Study aims to describe male first-line managers' experiences of their work situation in elderly care. First-line managers' work is challenging. However, less attention has been paid to male managers' work situation in health care. Knowledge is needed to empower male managers. Method Fourteen male first-line managers were interviewed. The interview text was subjected to qualitative content analysis. Result Work situations were described as complex and challenging; challenges were the driving force. They talked about 'Being on one's own but not feeling left alone', 'Having freedom within set boundaries', 'Feeling a sense of satisfaction and stimulation', 'Feeling a sense of frustration' and 'having a feeling of dejection and resignation'. Conclusion although the male managers' report deficiencies in the support structure, they largely experience their work as a positive challenge. Implications for nursing management to meet increasing challenges, male first-line managers need better access to supportive structural conditions. Better access to resources is needed in particular, allowing managers to be more visible for staff and to work with development and quality issues instead of administrative tasks. Regarding organizational changes and the scrutiny of management and the media, they lack and thus need support and information from superiors.

Studies conducted in health-care settings (Patrick & Laschinger 2006, Laschinger et al. 2007) have sup-ported Kanter's theory of structural empowerment and its applicability to managers' well-being. However, we have only found one study on FLMs' work situation looking at structural and psychological empowerment in municipal elderly care (Abdelrazek et al. 2010). Thus, our knowledge base is limited regarding the perspective of FLMs in elderly care, particularly as concerns empowerment. Moreover, there is little previous research (Finegan & Laschinger 2001) describing empowerment from a token perspective in a health care context. To our knowledge, there is no previous research exploring male FLMs' work situations in elderly care or in relation to the token perspective. Therefore, a deeper understanding and more knowledge are needed if we are to empower male FLMs working in elderly.

Study Power and empowerment in nursing: three theoretical approaches Definitions and uses of the concept of empowerment are wide-ranging: the term has been used to describe the essence of human existence and development, but also aspects of organizational effectiveness and quality. The empowerment ideology is rooted in social action where empowerment was associated with community interests and with attempts to increase the power and influence of oppressed groups (such as workers, women and ethnic minorities). Later, there was also growing recognition of the importance of the individual's characteristics and actions. Based on a review of the literature, this paper explores the uses of the empowerment concept as a framework for nurses' professional growth and development. Given the complexity of the concept, it is vital to understand the underlying philosophy before moving on to define its substance. The articles reviewed were classified into three groups on the basis of their theoretical orientation: critical social theory, organization theory and social psychological theory. Empowerment seems likely to provide for an umbrella concept of professional development in nursing (Patrick, 2006).

A study Aims to examine the influence of senior nurse leadership practices on middle and first-line nurse managers experiences of empowerment and organizational support and ultimately on their perceptions of patient care quality and turnover intentions. Background empowering leadership has played an important role in staff nurse retention but there is limited research to explain the mechanisms by which leader-ship influences nurse managers' turnover intentions. Methods this study was a secondary analysis of data collected using non-experimental, predictive mailed survey design. Data from 231 middle and 788 first-line Canadian acute care managers was used to test the hypothesized model using path analysis in each group. Results The results showed an adequate fit of the hypothesized model in both groups but with an added path between leadership practices and support in the middle line group. Conclusions Transformational leadership practices of senior nurses empower middle- and first-line nurse managers, leading to increased perceptions of organizational support, quality care and decreased intent to leave.

Studies showed a Positive relationships have been reported between Kouzes and Posner (1987) leadership practices and staff nurse empowerment (Patrick et al. 2011), organizational commitment, employee job satisfaction, productivity and decreased staff turnover (McNeese-Smith 1995, Chiok Foong Loke 2001, Houser 2003). Patrick et al. (2011) found that managers leadership practices had a significant positive direct effect on nurses structural empowerment (b = 0.69, P < 0.001). McNeese-Smith (1999) showed a relationship between staff nurses perceptions of their nurse manager s use of Kouzes and Posner s (1987) five leadership practices and job satisfaction. Better patient care outcomes closely linked to quality of care such as patient satisfaction, adverse events, and complications have also been associated with positive nursing leadership practices (Wong & Cummings 2007). To our knowledge, the impact of senior nurse leader s use of Kouzes and Posner s (1987) leadership practices on nurse managers perceived organizational support (POS), intent to leave or perceptions of quality care has not been studied.

Importantly, Kouzes and Posner s five leadership practices are teachable and can be adopted by nursing leaders through professional development programs. Recently, a study established that aspiring nurse leaders who underwent a 5-day leadership intervention program based on Kouzes and Posner s leadership model demonstrated an increase in the use of all five leadership practices (Tourangeau et al. 2003). Exemplary leadership practices were also used as the framework for a highly successful 4-year leadership development program for charge nurses in Colorado (Krugman & Smith 2003).

Study of how nurse managers' work-related empowerment has been investigated, in order to determine the level and relationships of empowerment among them. A systematic review was carried out, and a literature search was conducted with certain electronic databases for the period 1990–2009, using the main key words in various combinations. Only nine empirical studies in English were selected for review, in accordance with the requirements for the methodological quality and inclusion criteria. The most common type of study design was a descriptive survey (n = 5), and included various questionnaires, scales, and interviews. Nurse Managers' structural, psychological, and work empowerment was found to be high or moderately high. The empowerment of nurse managers correlated positively with job satisfaction, perceived organizational support, role satisfaction, and managerial self-efficacy, and correlated negatively with emotional exhaustion and own health outcomes. Different theoretical approaches ensure a clear understanding of empowerment, but difficulties arise when the findings are synthesized across studies and settings because of the different theoretical frameworks used to conceptualize empowerment (Tourangeau, 2003).

Further study conducted for purpose to find out how nurse managers' work-related empowerment has been investigated, in order to determine the level and relationships of empowerment in nurse managers. The review was guided by the following research questions:

(i)   What were the methodological characteristics of the empirical studies on the topic?

(ii)   How were nurse managers empowered?

(iii)   Which factors were connected to the nurse managers' empowerment?

No clear definition of empowerment was articulated in the empirical studies, with the exception of more recent studies drawing on specific theoretical frameworks of empowerment. It should be noted that the most frequently reported framework of empowerment was that proposed by Kanter (n = 5), representing the approach of organizational and management theories. Kanter's view of structural empowerment consists of organizational structures, access to which is influenced by the formal and informal power systems within the organization. (Tourangeau, 2003).

Structural empowerment had been measured by the CWEQ, with higher scores indicating higher empowerment. Furthermore, psychological empowerment, an alternative view of empowerment, is defined as a psychological response to an empowered work environment (n = 4), and reflects the approach of social psychological theories. It has been measured with Spreitzer's Psychological Empowerment Scale, composed of four dimensions: meaning, competence, self-determination, and impact. Higher scores reflect higher psychological empowerment. In addition, one study defined work-related empowerment as a process whereby the individual feels confident that he/she can act and successfully execute a certain kind of action, representing the empowerment as three unique but interdependent dimensions (verbal, behavioral, and outcome empowerment). This kind of empowerment is connected to the psychological approach, and is measured by the Work Empowerment Questionnaire by Irvine et al. (Suominen et al., 2005), with higher scores indicating higher levels of verbal, behavioral, and outcome empowerment. Moreover, Jooste (2000) created her own definition and structure of empowerment for investigation, emphasizing that it is a nurse leader's own characteristics, management skills, and responsibilities, which are in interaction with management structures, such as motivation, participation in decision-making, and power sharing. Unfortunately, Jooste (1997; 2000) did not present the measurement scores for her instrument. Kuokkanen and Leino-Kilpi (2001) utilized a semi structured interview to collect data on nurse managers' perceived empowerment.

In the majority of the empirical studies reviewed in this paper, nurse managers' structural, psychological, and work empowerment was reported to be high or of moderate level, the higher the nurse manager's position in the healthcare organization, the more empowering characteristics can be associated with her/him. Nurse Managers' empowerment scores were higher than staff nurses' scores, suggesting that they perceived themselves to have greater access to empowering structures. Furthermore, access to power is greater at higher hierarchical levels of an organization; therefore, middle managers were significantly more empowered than first-line managers (Goddard & Laschinger2015).

How to enhance manager power?

1) Expand personal resources. Power and energy go hand in hand. Effective leaders take sufficient time to unwind, reflect, rest, and have fun when they feel tired .

2) Present a powerful picture to other. How individuals look, act, and talk influence whether other view them as powerful or powerless. The nurse who stands tall and is poised, assertive,

articulate, and well-groomed presents a picture of personal control and power .

3) Pay the entry fee. Newcomers who stand out and appear powerful are those who do more, work harder, and contribute to the organization. They are not clock watchers or "nine-to-fivers." They attend meetings and in services, do committee work, and take their share of night shifts and weekend and holiday assignments without complaining .

4) Determine the powerful in the organization. Understanding and working within the formal and informal power structures are necessary. Individuals must be cognizant of their limitations and seek counsel appropriately .

5) Learn the language and symbols of the organization .

6) Learn how to use the organization's priorities. Every group has its own goals and priorities for achieving those goals. To build a power base must be cognizant of organizational goals and use those priorities and goals to meet management needs .

7) Increase professional skills and knowledge.

8) Maintain a broad vision. People without vision become very powerful .

9) Be flexible. Anyone wishing to acquire power should develop a reputation as someone who can compromise.

Keys to Employee Empowerment

a. Involve your employees in the decision-making process. In many companies, important decisions regarding the business come from senior level management. While this is a prudent approach to ensure the company's overall success, many times these decisions directly affect the lower level employees. If the final outcome of the decision is a less than positive one for the employees, often they feel mistreated and unappreciated. You can keep morale high and empower your employees by involving them in business decisions that directly relate to them .

b. Involve your employees in the planning process. In corporations across, self-directed work teams meet regularly to brainstorm ways to save money and work more efficiently. They take an active role in the goal-setting and planning process for the company as well as themselves. By doing so, they are not only becoming empowered, but they are also learning their own capabilities and limitations.

c. Offer praise freely. In order to empower your employees and develop a sense of competence and confidence, praise them for their performance, encourage them to take risks and acknowledge their accomplishments, no matter how small. This is particularly important when employees are beginning to work on new tasks or when an employee moves into a new job description.

d. Provide continual training and support. If you want your employees to feel confident of their abilities, they need the proper job training and support. Many companies offer an introductory training period when a new employee comes aboard, but then that training stop after a certain length of time. Unfortunately.

Notes

References:

- Abdelrazek F., Skytt B., Aly M., El-Sabour M.A., Ibrahim N. & Engstrom M. (2010) Leadership and management skills of first-line managers of elderly care and their work environ-ment. Journal of Nursing Management 18, 736–745.
- Advisory Committee on Health Human Resources, Ottawa. Chiok Foong Loke J. (2001) Leadership behaviours: effects on job satisfaction, productivity and organizational commitment. Journal of Nursing Management 9, 191–204.
- Aiken L.H., Clarke S.P. & Sloane D.M. (2002) Hospital staffing, organization, and quality of care: cross-national findings. International Journal for Quality in Health Care 14 (1), 5–13.
- Aitken LA, Sermens W, Van den Heede K, Sloane DM, Busse R, Bruyneel L, Raf-ferty AM, Griffiths P, Moreno-Casbas MT, Tishelman C, Scott A, Brzostek T, Kinnunen J, Schwendimann R, Heinta M, Zikos D, Sjetne IS, Smith HL & Kutney-Lee A (2012) Patient safety, satisfaction and quality of care: cross-sectional surveys on nurses and patients in 12 countries in Europe and the United States. British Medical Journal 344, e1717.
- Anderson J. (1996) Empowering patients: issues and strategies.Social Science and Medicine 43, 697±705.
- Andrews, G . (1994) . Mistrust . the hidden obstacle to empowerment. HR Magazine . 39(9), 66-70.
- Armellino D, Griffin M & Fitzpatrick J (2010) Structural empowerment and patient safety culture among registered nurses working in adult critical care units. Journal of Nursing Management 18, 796–803.
- Arneson H, Ekberg K. Measuring empowerment in working life: a review. Work 2006; 26: 37–46.
- Backer B., Costello-Nickitas D. & Mason-Adler M. (1994) Nurses' experiences of empowerment in the workplace: a qualitative study. Journal of the New York State Nurses Association 25, 4±7.

- Bandura A. (1977) Self-ef®cacy: toward a unifying theory of behavioral change. Psychological Review 84, 191±215.
- Bandura A. (1978) The self system in reciprocal determinism. American Psychologist 33, 344±369.
- Blegen, M . A . (1993). Nurses' job satisfaction : A meta-analysis of related variables. Nursing
- Bradshaw A (2010) Is the ward sister role still relevant to the quality of patient care? A critical examination of the ward sister role past and present. Jour-nal of Clinical Nursing 19, 3555–3563.
- Buchanan, B . (1974) . Building organizational commitment : The socialization of managers in work organizations . Administrative Science Quarterly, 19, 533-546.
- Canadian Institute for Health Information. (2006) Workforce Trends of Registered Nurses in Canada, 2006. Registered Nurses Database, Ottawa.
- Canadian Nursing Advisory Committee. (2002) Our Health, Our Future: Creating Quality Workplaces for Canadian Nurses.
- Cangemi, J . P ., Rice, J ., & Kowalski, C . J . (1989) . The development, decline and renewal of trust in an organization: Some observations . Organizational Development Journal (Winter), 2-9.
- Chandler G. (1991) Creating an environment to empower nurses. Nursing Management 22, 20±23.
- Chandler G. (1992) The source and process of empowerment.Nursing Administration Quarterly 16, 65±71.
- Cheek J. & Porter S. (1997) Reviewing Foucault: possibilities and problems for nursing and health care. Nursing Inquiry 4, 108±119.
- Cicolini G, Comparicini D & Simonetti V (2014) Workplace empowerment and nurses job satisfaction: a systematic review. Journal of Nursing Manage-ment 22, 855–871.
- Clifford P.G. (1992) The myth of empowerment. Nursing Administration Quarterly 16, 1±5.

- Clutterbuck D. (2014) The Power of Empowerment. Release the Hidden Talents of Your Employees. Kogan Page, London.
- Conger J & Kanugo B (1988) The empow-erment process: integrating theory and practice. Academy of Management Review 13, 471–482.
- Conger J. & Kanungo R. (1988) The empowerment process: integrating theory and practice. Academy of Management Review 13, 471±482.
- Daley, D . M . (1991) . Management practices and the uninvolved manager : The effect of supervisory attitudes on perceptions of organizational trust and change orientation . Public Personnel Management, 20(1), 101-113.
- Doering L. (1992) Power and knowledge in nursing: a feminist poststructuralist view. Advances in Nursing Science 14, 24±33.
- Elhabashy, S., BRONCHIAL HYGIENE THERAPY: Modalities & Techniques , , New Jersey, Princeton University Press, 2016 .
- Elhabashy, S., Cardio-Thoracic Injury, Essentials All Critical Care Nurses Need To Know, U.S.A, Elseveir (MOSBY), 2015. the_final_book.pdf
- Elhabashy, S., Clinical Alarms Hazards and Management at Critical Care Settings, Newyork, LULU Publishing Com., 2015. 16932240_cover.pdf
- Elhabashy, S., factors affecting validity of arterial blood gases results, Germany, LAMBERT Academic Publishing, 2013 .
- Elhabashy, S., Formulate Consequential Student Learning Outcomes, USA, Johns Hopkins University Press, 2017 .
- Elhabashy, S., R. Elkodoos, S. Gebril, B. Osman, M. Elsawy, S. Hamad, N. Kasem, M. Mostafa, A. Mahrous, and M. AbouZead, "A Review of Medical Devices-Related Pressure Ulcers", International Journal of Biology, Pharmacy and Allied Sciences , vol. 7, issue 6, pp. 1133-1146, 2018. httpsdoi.org10.31032ijbpas20187.6.4473.pdf
- Eschleman, K. J., Bowling, N. A., & Alarcon, G. (2010). A meta-analytic examination of hardiness. International Journal of Stress Management, 17, 277–307.

- Foucault M. (2012) The Archaeology of Knowledge. Tavistock, London.
- Freire P. (1972) Pedagogy of the Oppressed. Penguin, Harmonds-worth.
- Fulton Y. (1997) Nurses' view on empowerment: a critical social theory perspective. Journal of Advanced Nursing 26, 529±536.
- Fulton Y. Nurses' views on empowerment: a critical social theory perspective. J. Adv. Nurs. 1997; 26: 529–536.
- Gaber Sh., Elhabashy S. "Applied Advocacy for Healthcare Professionals", Advances in Bioresearch, 4(6), 2018 pp. 167–175.
- Gaber, S., and N. Fekry, ARE Emergency Nurses Well Prepared for Disaster Management? , Germany, LAP LAMBERT Academic Publisher , 2015. 978-3-659-67138-8_coverpreview-3.pdf
- Gaber, S., Disaster management at Health Care Settings Comprehensive assessment and effective mitigation, , U.S.A, LULU. Press, 2015. final_shreen_book-1.pdf
- Gaber, S., F. abed, and I. Said, "Effect of a Developed Evidence-Based Discharge Protocol on Cancer Colon Patients Satisfaction ", Impact Journals, vol. 4, issue 9, pp. 167-176, 2016.
- Gaber, S., Global Citizenship in Nursing, USA, Stanford University Press, 2017.
- Gaber, S., Nursing as a Profession and Patient Leading, Guidance, & Support, USA, Yale University Press, 2016.
- Gibson CH. A concept analysis of empowerment. J. Adv. Nurs. 2017; 16: 354–361.
- Gilbert T. (1995) Nursing: empowerment and the problem of power. Journal of Advanced Nursing 21, 865±871.
- Gilbert T. Nursing: empowerment and the problem of power. J. Adv. Nurs. 1995; 21: 865–871.
- Goddard P. & Laschinger H.K.S. (2015) Evaluating physical therapists perception of empowerment using Kanter s theory of structural power in organizations. Physical Therapy 81 (12), 1880–1888.

- Habermas J. (1978) Knowledge and Human Interests 2nd edn. Heineman, London.
- Habermas J. (1979) Communication and the Evolution of Society. Beacon Press, Boston.
- Hagerman H, Engstrom M, Hagstrom E, Waldstein B & Skytt B (2015) Male first line managers experiences of the work situation in elderly care: an empowerment perspective. Journal of Nursing Management 23, 695–704.
- Hakansson C. & Ahlborg G. Jr (2010) Perceptions of employ- ment, domestic work, and leisure as predictors of health among women and men. Journal of Occupational Science 17, 150–157.
- Harden J. (1996) Enlightenment, empowerment and emancipa- tion: the case for critical pedagogy in nurse education. Nurse Education Today 16, 32±37.
- Hart, K . M ., Capps, H . R ., Cangemi, J. P ., & Caillouet, L . M . (1996) . Exploring organizational trust and its multiple dimensions : A case study of General Motors . Organization Development Journal (Summer), 31-39.
- Havens D.S. & Laschinger H.S. (1997) Creating the environment to support shared governance; Kanter s theory of power in organizations. Journal of Shared Governance 3 (1), 15–23.
- Heslop L. (1997) The (im)possibilities of poststructuralist and critical social nursing inquiry. Nursing Inquiry 4, 48±56.
- Hess R. (1984) Thoughts on Empowerment in Studies in Empow-erment. The Haworth Press, New York.
- Hewison A. (1995) Nurses' power in interactions with patients. Journal of Advanced Nursing 21, 75±82.
- http://hsb .baylor.edu/ramsower/ais.ac.96/papers/sethi.htm.
- https://doi.org/10.1037/a0020476 European Agency for Safety and Health at Work (2014). Calculating the cost of work-related stress and psychosocial risks. Luxembourg: Publications Office of the European Union. https://doi. org/10.2802/20493
- Huntington A., Gilmor J. & O'Connell A. (1996) Reforming the practice of nurses: decolonization or getting out from under. Journal of Advanced Nursing 24, 364±367.

- Hystad, S. W., Eid, J., & Brevik, J. I. (2011). Effects of psychological hardiness, job demands, and job control on sickness absence: A prospective study. Journal of Occupational Health Psychology, 16, 265–278. https://doi.org/10.1037/a0022904
- Irvine, D . M ., & Evans, M . G . (1995) . Job satisfaction and turnover among nurses : Integrating
- Ismail, M. S., A. Y. Zakaria, Taema, and S. Elhabashy, "ENDOTRACHEAL TUBE PRESSURE INJURY: NURSING PREVENTIVE MEASURES", IMPACT : International Journal of Research in Applied, Natural and Social Sciences (IMPACT : IJRANSS) , vol. 5, issue 10, pp. 9-16, 2017. 3abstract.docx
- Jooste K. A comparison of the viewpoints of different levels of nurse managers on empowerment in their workplace. Health SA Gesondheid. 2000; 5: 15–29.
- Jooste K. A model for the empowerment of nurses: a management perspective. Health SA Gesondheid. 1997; 2: 32–40.
- Judkins, S., Furlow, L., & Kendricks, T. (2007). A training program was conducted to reduce stress and enhance hardiness among nurses. Nursing Management, 14(7), 19–23. https://doi.org/10. 7748/nm.14.7.19.s14
- Kanter R.M. (1979) Power failure in management circuits. Harvard Business Review 4, 65±75.
- Kanter R.M. (1977) Men and Women of the Corporation. Basic Books, New York.
- Kanter R.M. (1979) Men and Women of the Corporation. Basic Books, New York, NY.
- Kanter RM (1977) Men and Women of the Corporation. Basic Books, New York, NY.
- Kanter RM (1979) Power failure in man-agement circuits. Harvard Business Review July–August: 65–75.
- Kanter RM (1993) Men and Women of the Corporation, 2nd edn. Basic Books, New York, NY.

- Kerfoot K. To empower or partner? The challenge for the nurse manager. Nurs. Econ. 1994; 12: 173–174.
- Kieffer C.H. (1984) Citizen empowerment: a developmental perspective. Prevention in Human Services 3, 9±36.
- Kirkpatrick B. (1992) Original Roget's Thesaurus: of English Words and Phrases. Longman Group, Harlow.
- Kluska K.M., Laschinger H.K.S. & Kerr M.S. (2004) Staff nurse empowerment and effort-reward imbalance. Canadian Journal of Nursing Leadership 17 (1), 112–128.
- Kobasa, S. C. (1979). Stress life events, personality, and health: An inquiry into hardiness. Journal of Personality and Social Psychology, 37, 1–11. https://doi.org/10.1037/0022-3514.37.1.1
- Kobasa, S. C., Maddi, S. R., & Kahn, S. (1982). Hardiness and health: A prospective study. Journal of Personality and Social Psychology, 42, 168–177. https://doi.org/10.1037//0022-3514.42.1.168
- Kouzes J.M & Posner B.Z. (1987) The Leadership Challenge:How to Get Extraordinary Things Done in Organizations.Josey-Bass, San Francisco, CA.
- Kramer & T . R . Tyler The decisions to trust as a social decision . In: (Eds), Trust in Organizations : Frontiers of Theory and Research (pp. 357-389) . Thousand Oaks, CAL: Sage Publications.
- Kramer, M ., & Schmalenberg, C . (1993) . Learning from success : Autonomy and empowerment.
- Krugman M. & Smith V. (2003) Charge nurse leadership development and evaluation. Journal of Nursing Administration 33, 284–292.
- Kuokkanen L, Leino-Kilpi H. Power and empowerment in nursing: three theoretical approaches. J. Adv. Nurs. 2000; 31: 235–241.
- Kuokkanen L, Leino-Kilpi H. The qualities of an empowered nurse and the factors involved. J. Nurs. Manag. 2001; 9: 273–280.

- Laschinger H. (1996) A theoretical approach to studying work empowerment in nursing: a review of studies testing Kanter's theory of structural power in organizations. Nursing Administration Quarterly 20, 25±41.
- Laschinger H., Sabiston J. & Kutszher L. (1997) Empowerment and staff nurse decision involvement in nursing work environments: testing Kanter's theory of structural power in organizations. Research in Nursing and Health 20, 341±352.
- Laschinger H.K.S. & Shamian J. (2016) Staff nurses and nurse managers perceptions of job-related empowerment and managerial self-efficacy. Journal of Nursing Administration 24 (10), 38–47.
- Laschinger H.K.S. (2004) Hospital nurses perceptions of respect and organizational justice. Journal of Nursing Administration 34 (7/8), 354–364.
- Laschinger H.K.S. (2008a) Effect of empowerment on professional practice environments, work satisfaction, and patient care quality: further testing the nursing work life model. Journal of Nursing Care Quality 23 (4), 322–330.
- Laschinger H.K.S. (2008b) UWO Workplace Empowerment Program. Available at: http://publish.uwo.ca/~hkl/, accessed 16 December 2008.
- Laschinger H.K.S., Almost J. & Tuer Hodes D. (2003a) Workplace empowerment and magnet hospital characteristics: making the link. Journal of Nursing Administration 33 (7/8), 410–422.
- Laschinger H.K.S., Almost J., Purdy N. & Kim J. (2010) Predictors of nurse managers health in Canadian restructured healthcare settings. Nursing Leadership 17 (4), 88–105.
- Laschinger H.K.S., Finegan J. & Wilk P. (2009) Context matters. The impact of unit leadership and empowerment on nurses organizational commitment. Journal of Nursing Administration 39 (5), 228–235.
- Laschinger H.K.S., Finegan J., Shamian J. & Wilk P. (2001) Impact of structural and psychological empowerment on job strain

in nursing work settings: expanding Kanter s model. Journal of Nursing Administration 31 (5), 260–272.

- Laschinger H.K.S., Finegan J., Shamian J. & Wilk P. (2003b) Workplace empowerment as a predictor of nurse burnout in restructured healthcare settings. Hospital Quarterly 6 (4), 2–11.
- Laschinger H.K.S., Leiter M., Day A. & Gilin D. (2009) Work-place empowerment, incivility, and burnout: impact on staff nurse recruitment and retention outcomes. Journal of Nursing Management 17, 302–311.
- Laschinger H.K.S., Purdy N. & Almost J. (2007) The impact of leader-member exchange quality, empowerment, and core self-evaluation on nurse manager s job satisfaction. Journal of Nursing Administration. 37 (3), 221–229.
- Laschinger HKS, Purdy N, Almost J. The impact of leader-member exchange quality, empowerment, and core self-evaluation on nurse manager's job satisfaction. J. Nurs. Adm. 2007; 37: 221–229.
- Laschinger HSK (2010) Positive working relationships matter for better nurse and patient outcomes. Journal of Nursing Management 18, 875–877.
- Laschinger HSK, Day A, Leiter MP, Gilin-Oore D & Mackinnon SP (2012) Building empowering work environ-ments that foster civility and organisa-tional trust: testing an intervention. Nursing Research 61, 316–325.
- Lee H. & Cummings G.G. (2008) Factors influencing job satis-faction of front line nurse managers: a systematic review. Journal of Nursing Management 16, 768–783.
- Lim, J., Bogossian, F., & Ahern, K. (2010). Stress and coping in Australian nurses: A systematic review. International Nursing Review, 57(1), 22–31. https://doi.org/10.1111/j.1466-7657. 2009.00765.x
- Lundberg U. & Frankenhaeuser M. (2018) Stress and workload of men and women in high-ranking positions. Journal of Occupational Health Psychology 4, 142–151.

- Lundqvist, D., Reineholm, C., Gustavsson, M., & Ekberg, K. (2013). Investigating work conditions and burnout at three hierarchical levels. Journal of Occupational and Environmental Medicine, 55, 1157–1163. https://doi.org/10.1097/jom.0b013e31829b27df
- Macphee M., Skelton-Green J., Bouthillette F. & Suryaprakash N. (2012) An empowerment framework for nursing leader-ship development: supporting evidence. Journal of Advanced Nursing 68, 159–169.
- Manojlavich M (2005) The effect of nurs-ing leadership on hospital nurses pro-fessional practice behaviours. Journal of Nursing Administration 35, 366– 374.
- Marin-Garcia, J. A., & Bonavia, T. (2015). Relationship between employee involvement and lean manufacturing and its effect on performance in a rigid continuous process industry. International Journal of Production Research, 53, 3260–3275. https://doi.org/10.1080/ 00207543.2014.975852
- Marquis BL, Huston CJ. Leadership Roles and Management Func-tions in Nursing: Theory and Application. Philadelphia, PA: Lip-pincott Williams & Wilkins, 2000.
- Maslach, C., Schaufeli, W. B., & Leiter, M. P. (2001). Job burnout. Annual Review of Psychology, 52, 397–422. https://doi.org/10.1146/annurev.psych.52.1.397
- Maynard, M. T., Gilson, L. L., & Mathieu, J. E. (2012). Empowerment fad or fab? A multilevel review of the past two decades of research. Journal of Management, 38, 1231–1281. https://doi.org/10. 1177/0149206312438773
- McDaniel, C ., & Stumpf, L . (1993) . The organizational culture : Implications for nursing science. Journal of Nursing Administration, 23(4), 54-60.
- McKinney, R., McMahon, M., & Walsh, P. (2013). Danger in the middle: Why midlevel managers aren't ready to lead. Retrieved from http://www.harvardbusiness.org/sites/default/files/PDF/ 17807_CL_MiddleManagers_White_Paper_March2013.pdf

- McNay L. (2017) Foucault, a Critical Introduction. Polity Press, London.
- McNeese-Smith D. (1995) Job satisfaction, productivity, and organizational commitment: the result of leadership. Journal of Nursing Administration 25 (9), 17–26.
- Meyer, J . P ., & Allen, N. J . (1997). Commitment in the workplace : Theory, research, and application . Thousand Oaks, CA : Sage Publications.
- Meyer, J . P ., Bobocel, D. R ., & Allen, N . J . (1991) . Development of organizational commitment during the first year of employment: A longitudinal study of pre- and post-entry influences . Journal of Management, 17(4), 717-733.
- Meyer, J . P ., Irving, P. G ., & Allen, N . J . (1998) . Examination of the combined effects of work values and early work experiences on organizational commitment . Journal of Organizational Behaviour, 19, 29-52.
- Meyer, J . P., Allen, N . J ., & Smith, C . A . (1993) . Commitment to organizations and occupa-tions: Extension and test of a three-component conceptualization . Journal of Applied Psychology, 78(4), 538-551.
- Mishra, A . K., & Spreitzer, G . M. (1998) . Explaining how survivors respond to downsizing : The roles of trust, empowerment, justice and work redesign . Academy of Management Review, 23(3), 567-588.
- Nursing Management, 25(5), 58-64.
- Oshry B (2007) In the Middle. Power & Systems Incorporated, Bos-ton, MA.
- Oshry B (2016) Seeing Systems: Unlocking the Mysteries of Organisational Life, 2nd edn. Berrett–Koehler Publishers, San Francisco, CA.
- Ouchi, W . G . (1981) . Theory Z. New York: Addison-Wesley.
- Parker B. & McFarlane J. (1991) Feminist theory and nursing: an empowerment model for research. Advances in Nursing Science 13, 59±76.

- Partridge E. (1966) Origins, a Short Etymological Dictionary of Modern English. Routledge and Kegan Paul, London.
- Patrick A & Laschinger HSK (2006) The effect of structural empowerment and perceived organizational support on middle level nurse managers' role sat-isfaction. Journal of Nursing Manage-ment 14, 13–22.
- Patrick A, Laschinger HSK, Wong C & Finegan J (2007) Developing and test-ing a new measure of staff nurse clini-cal leadership: the clinical leadership survey. Journal of Nursing Manage-ment 19, 449–460.
- Patrick A., Laschinger H.K.S., Wong C. & Finegan J. (2011) Developing and testing a new measure of staff nurse clinical leadership: the clinical leadership survey. Journal of Nursing Management 19, 449–460.
- Podsakoff, P . M ., MacKenzie, S . B., & Bommer, W. H . (1996) . Transformational leadership behaviours and substitutes for leadership as determinants of employee satisfaction, commitment, trust, and organizational citizenship behaviors . Journal of Management.
- Procter S, Currie G, Orme H. Limits to employee empowerment in the UK NHS: locality managers in a community health trust. J. Manag. Med. 2015; 13: 405–420.
- Purdy N, Laschinger HSK, Finegan J, Kerr M & Olivera F (2010) Effects of work environment on nurse and patient out-comes. Journal of Nursing Manage-ment 18, 901–913.
- Rafael A. (1996) Power and caring: a dialectic in nursing. Advances in Nursing Science 19, 3±17.
- Rappaport J. (1984) Studies in empowerment: introduction to the issue. Prevention in Human Services 3, 1±7.
- Reay T., Golden-Biddle K. & Germann K. (2003) Challenges and leadership strategies for managers of nurse practitioners. Journal of Nursing Management 11, 396–403.
- Regan L & Rodriguez L (2011) Nurse empowerment from a middle manage-ment perspective: nurse managers' and assistant

nurse managers' workplace empowerment views. The
Permanente Journal 15, e101–e107.
- Regan L, Lachinger HSK & Wong CA (2015) The influence of
empowerment, authentic leadership and professional practice
environments on nurses per-ceived interprofessional
collaboration. Journal of Nursing Management 24, E54–E61.
- Regan L.C. & Rodriguez L. (2011) Nurse empowerment from a
middle-management perspective: nurse managers and assistant
nurse managers workplace empowerment views. Permanente
Journal 15 (1), e101–e107. Available at: http://www.theper-
manentejournal.org/files/Winter2011PDFS/NurseManagers.pdf,
accessed 05 May 2011.
- Research, 42(1), 36-41. Downloaded by EKB Data Center At
10:43 03 December 2018 (PT(HEATHER K . SPENCE
LASCHINET AL.
- Rhoades L. & Eisenberger R. (2002) Perceived organizational
support: a review of the literature. Journal of Applied Psychology 87 (4), 698–714.
- Robbins A. (1986) Unlimited Power. Ballantine Books, New
York. Roberts S.T. (1983) Oppressed group behavior:
implications for nursing. Advances in Nursing Science 5, 21±30.
- Rodwell C. (1996) An analysis of the concept of
empowerment.Journal of Advanced Nursing 23, 305±313.
- Sabiston J. & Laschinger H. (2015) Staff nurse work
empowerment and perceived autonomy. Testing Kanter's theory
of structural power in organizations. Journal of Nursing
Administration 25, 42±50.
- Salanova, M., Schaufeli, W. B., Llorens, S., Peiro, J. M., & Grau,
R. (2000). Desde el "burnout" al "engagement": ¿una nueva
perspectiva? [From the "burnout" to "engagement": A new
perspective?]. Revista de Psicologia del Trabajo y de las
Organizaciones, 16(2), 117–134.
- Sarmiento T.P., Laschinger H.K.S. & Iwasiw C. (2004) Nurse
educators workplace empowerment, burnout, and job satis-

faction: testing Kanter s theory. Journal of Advanced Nursing 46 (2), 134–143.
- Schlesinger A & Oshry B (1984) Quality of work life and the manager: muddle in the middle. Organizational Dynam-ics 13, 5–19.
- Sethi, V ., Meinert, D ., King, R . C ., & Sethi, V . (1996) . The multidimensional nature of organizational commitment among information systems personnel . [On-line] . Available:
- Sheldon L. & Parker P. (1997) The power to lead. Nursing Management 4, 8±9.
- Shirey M.R., Ebright P.R. & McDaniel A.M. (2008) Sleepless in America: nurse managers cope with stress and complexity. Journal of Nursing Administration 38, 125–131.
- Simpson J. & Bradley W. (1989) Oxford English Dictionary 2nd edn, Volume V. Clarendon Press, Oxford.
- Spence HK, Lasant KS, Wong CA, Grau AL, Read EA & Stam LMP (2012) The influence of leadership practices and empowerment on Canadian nurse manager outcomes. Journal of Nursing Management 20, 877–888.
- Spinks J (2010) Do ward sisters and charge nurses have the authority to do their job? Nursing Management 17, 20–22.
- Spreitzer G. (1995) Psychological empowerment in the work-place: dimensions, measurement and validation. Academy of Management Journal 38 (5), 1442–1462.
- Spreitzer G. (1996) Social structural characteristics of psycholog-ical empowerment. Academy of Management Journal 39, 483±504.
- Spreitzer G.M. (1995) Psychological empowerment in the work-place: dimensions, measurement, and validation. The Acad-emy of Management Journal 38, 1442–1465.
- Spreitzer, G. M. (2008). Taking stock: A review of more than twenty years of research on empowerment at work. In C. Cooper & J. Barling (Eds.), Handbook of organisational behavior (pp. 54–73). Thousand Oaks, CA: Sage.

- Suominen T, Savikko N, Puuka P, Doran D & Leino-Klipi H (2005) Work empowerment as experienced by head nurses. Journal of Nursing Manage-ment 13, 147–153.
- Swedish National Board of Health and Welfare (2011) (in Swedish) L€agesrapport 2011: h€also- och sjukvard och socialtj€anst. Stockholm. Swedish National Board of Health and Welfare. Available at: http://www.socialstyrelsen.se/Lists/ Artikelkatalog/Attachments/18229/2011-2-1.pdf, accessed 15 April 2013.
- Tansky J.W. & Cohen D.J. (2001) The relationship between organizational support, employee development, and organiza-tional commitment: an empirical study. Human Resource Development Quarterly 12 (3), 285–300.
- Tengelin E., Arman R., Wikstrom E. & Dellve L. (2011) Regu-lating time commitments in healthcare organizations: manag-ers' boundary approaches at work and in life. Journal of Health Organization and Management 25, 578–599.
- Thomas K. & Velthouse B. (1990) Cognitive elements of empow-erment: an `interpretive' model of intrinsic task motivation. Academy of Management Review 15, 666±681.
- Thomas K.W. & Velthouse B.A. (1990) Cognitive elements of empowerment: an interpretive model of empowerment in leadership. Academy of Management Review 15, 666–681.
- Tones K. (1994) Health promotion, empowerment and action competence. In Action and Action Competence as Key Concepts in Critical Pedagogy (Jensen B. & Schnack K. eds), Royal Danish School of Educational Studies, Copenhagen, pp. 163±183.
- Tourangeau A.E., Lemonde M., Luba M., Dakers D. & Alksnis C. (2003) Evaluation of a leadership development intervention.
- Tyler, T. R ., & Degoey, P . (1996) . Trust in organizational authorities: The influence of motiveattributions on willingness to accept decisions . In: R . M . Kramer & T. R. Tyler (Eds‹(Trust in Organizations : Frontiers of Theory and Research (pp . 331-356) . Thousand Oaks.‹

- Van Bogaert, P., Kowalski, C., Weeks, S. M., Van Heusden, D., & Clarke, S. P. (2013). The relationship between nurse practice environment, nurse work characteristics, burnout and job outcome and
- Vazquez-Bustelo, D., & Avella, L. (2017). The effectiveness of high-involvement work practices in manufacturing firms: Does context matter? Journal of Management & Organization. Advance online publication. https://doi.org/10.1017/jmo.2016.69
- Vogt J. & Murrell K. (1990) Empowerment in Organizations. How to Spark Exceptional Performance. Pfeiffer & Company, San Diego.
- Wagner JIJ, Cummings G, Smith DL, Olson J, Anderson L & Warren S (2012) The relationship between struc-tural empowerment and psychological empowerment for nurses: a systematic review. Journal of Nursing Manage-ment 18, 448–462.
- Wang X, Kunaviktikal W & Wichaikhum O (2014) Work empowerment and burnout amongst registered nurses in
- Ward D. & Mullender A. (1991) Empowerment and oppression: an indissoluble pairing for contemporary social work. Critical Social Policy 11, 21±30.
- Weissman, C ., & Nathanson, C. (1985) . Professional satisfaction and client outcomes . Medical Care, 23, 1179-1193.
- Whitney, J . (1994) . The trust factor . New York: McGraw Hill The Impact of Workplace Empowerment   85.
- Williams C.L. (1992) The glass escalator: hidden advantages for men in the 'female' professions. Social Problems 39, 253–267.
- Wood, S., Van Veldhoven, M., Croon, M., & de Menezes, L. M. (2012). Enriched job design, high involvement management and organisational performance: The mediating roles of job satisfaction and well-being. Human Relations, 65, 419–445. https://doi.org/10.1177/0018726711432476/.
- Young-Ritchie C, Laschinger HSK & Wong C (2009) The effects of emo-tionally intelligent leadership beha-viour on emergency

staff nurses' workplace empowerment and organi-zational commitment. Nursing Leader-ship 22‹

- Zakaria, A. Y., Taema, K. M., Ismael, M. S., Elhabashy, S. (2018). Impact of a suggested nursing protocol on the occurrence of medical Device-related pressure ulcers in Critically Ill Patients. Central European Journal of Nursing and Midwifery, 9(4), 924-931 .doi: 10.15452/CEJNM.2018.09.0025

www.ingramcontent.com/pod-product-compliance
Lightning Source LLC
Chambersburg PA
CBHW030907180526
45163CB00004B/1749